Nature, to my mind, gave man three materials to serve him in the course of his life: earth, in which to grow food; wood, from which to fashion furniture; and stone, of which to build his home.
Frazier Peters

The Owner Builder's Guide to STONE MASONRY

Ken Kern • Steve Magers • Lou Penfield

Published by CHARLES SCRIBNER'S SONS NEW YORK

This edition published by Charles Scribner's Sons 1980

1 3 5 7 9 11 13 15 17 19 H/P 20 18 16 14 12 10 8 6 4 2

Printed in the United States of America
Library of Congress Catalog Card Number 80-52494
ISBN: 0-684-15288-6

Contents

Introduction

A book about stone masonry should be written by a stonemason. Even the most perceptive observer, however well grounded in the techniques of stonework, would be at a loss to capture for readers the essence of this activity, the spiritedness realized as one stone is laid against another. The subject is one topic requiring the direct involvement of the artist-builder before verbal evaluation of the experience becomes meaningful for others.

We three authors have each experienced the animation associated with this skill: Lou is an artist and teacher who lives in Ohio and built his studio-in-the-woods of stone from nearby Chagrin River; Steve is a stonemason from Pittsboro, North Carolina, who specializes in innovative stonework; Ken is a west coast mason who also designs and writes about low-cost, owner-built housing. Together we have combined our regard for stonework with our various backgrounds and talents to acquaint the reader with the experience and practice of building with this unique material.

The purpose of this book is threefold. To begin, we wish to dispel as unfounded the historically-held myth that stonemasons are gifted with some special sense for pursuing this craft. On the contrary, as masons themselves facetiously declare, stone laying requires "little more than a strong back and a weak mind." This "little more" of which they jest is the basis of this book, with its step-by-step approach to building stone structures. We show the inexperienced builder how to "lay up" stone for various walls, how to "face" building framework and how to "cast" stone in a wall with a movable form.

A second purpose of this manual acquaints readers with the native properties and the availability of useable building stone. Next to earth there is no more universal nor less appreciated building resource than stone. This ideal substance may be found virtually everywhere, although it is usually moved aside or hauled away to prepare a pad for wood-framed or other kinds of construction.

An example of the utter disregard for building with stone is seen in the accompanying photo from a California newspaper (Fresno Bee, 7 April 1976). A Merced County supervisor and a local landowner appear to be inspecting a nondescript pile of stone. This material is, however, a perfect size, shape and quality for building purposes, but these two gentlemen are discussing ways to *bury* it! Fifty years of gold mining were required to extract these millions of tons of stone now piled on 13,000 acres. The plan is to invest $6 million (more than the amount of the gold originally mined here) in a dozen giant dredges, each designed to reclaim one acre a day by scooping up soil from a depth of twenty-five feet, burying the stone in the excavations and thereby reclaiming the land for agricultural purposes. Federal and county funds presumably will finance the ten-year project at an undisclosed cost.

For curiosity, one of the authors pursued this news item, uncovering the following facts: Merced is a relatively poor agricultural county. At the time the Bee article appeared 6,900 people in the area were unemployed. The housing allocation index showed 8,114 houses were classified as "primary blight," requiring immediate replacement or rehabilitation. An additional 4,558 houses were classified as "secondary blight" and about 1,000 houses were unsalvageable and destined for immediate demolition. What kind of imagination or mentality is necessary to bring 6,900 people needing work together with millions of tons of neatly piled stone, ultimately arriving at the equitable use of taxpayer funds for the solution to an urgent need for 14,000 houses?

Our third impetus for writing this book is to express the aesthetic satisfaction we three authors have experienced building with stone. It is not only possible for thousands of people to build thousands of urgently needed homes

but they can have a ball doing so! In this regard, one is reminded of the stonemason who was interviewed by Studs Terkel in his book, *Working.* In over one hundred interviews with workers in America today, the stonemason stands forth as *the only person* not dissatisfied with his work.

In discussion with Terkel, this mason begins by telling some of the history of stonework. How many workers today know - or give a damn - about the history, the traditions, or desciplines of their craft? As he talks, the mason described his thoughts and feelings for work with stone.

"Stone's my life," he says. "I daydream all the time, most times it's on stone. Oh, I'm gonna build me a stone cabin down on the Green River. I'm gonna build stone cabinets in the kitchen. That stone door's gonna be awful heavy and I don't know how to attach the hinges. I've got to figure out how to make a stone roof. That's the kind of thing. All my dreams, it seems like it's got to have a piece of rock mixed in it."*

*Terkel, Studs. *Working*, Pantheon Books, a Division of Random House, Inc., NY, 1972, p. 20.

The authors concur with these sentiments. Each in his own way has experienced similar feelings for work with stone. When building with this intriguing material one communes with antiquity at the same time that one builds with stability, permanence and feasibility. The primary cost of stonework is the satisfying labor that accompanies it.

We also know that these feelings need not be the exclusive province of the professional turned-on mason. Every person working with stone can acquire a high regard for building with this resource, but we do not wish to imply that other construction media fail to offer the owner-builder equal opportunity for joy and creativity - merely that other means of construction are not as unused, as unappreciated as stone masonry. In this circumstance, we seek to impress readers with the value of stonework and to urge upon them a willingness to attempt some building with this neglected material. We hope to guide the inexperienced builder in construction with stone.

North Fork CA
August 1976

Basics

Overview

This chimney was built over a hundred years ago on one end of a wooden farmhouse. Stone was used for construction because it was available. At that time brick had limited use because it was expensive and difficult to obtain. The wood-framed house is gone now but the chimney stands as a momument to its maker.

If you could walk closer to examine this chimney it would be apparent that mortar was not used to hold it together, only red clay mud. Most of the mud washed away years ago and what remains is a neatly stacked pile of stone. A clear look at this material reveals nothing special; it is not particularly square-just stone that was plowed from a field or dug up with well excavation. Since it is heavy and hard to transport it is unlikely there is a stone in this chimney which is not native to this place.

Looking still closer one sees what was done with this common building material. The mason obviously understood this medium; all stones in the chimney work together as a unit, each adding its part to make the whole. The vertical joints are crossed by stone, above and beneath. All stones sit firmly, weighting down those below and at the same time providing a flat, firm bed of support for those placed on top. There are no gapping holes between stones so that each seems to melt into the next. In fact, it seems that over the years they have grown together.

It is while looking at such work that one understands why stone masonry is seen by many with an air of mystical wonder. "How do those stones fit together?" "Why don't they topple?" Such comments indicate observers' assumptions that the mason must have known some secret process which kept the chimney from falling. In truth, masonry is as fundamental as gravity, nothing more. Once you are aware that a stone's tendency is to fall straight down you will know how to build a strong wall. Our book is merely an elaboration of this fact.

Gathering Stone

The process of building with stone begins with the selection and gathering of the material. By studying various walls, foundations and chimneys you will see that stone can be found in an endless variety of shapes, textures and sizes. Since the type of stone selected partially determines not only the way a structure will look but even the way it will be built, it is wise to scout around to discover as many sources as possible. Only then can you choose the type which best suits your purpose.

Many owner-builders prefer to use material from their own land. There are several reasons for this choice. Aesthetically, if the goal is to build a house that fits its surroundings, even "grows" from its surroundings, then it merits the use of stone from the land. For purely practical reasons, collecting material from one's land is often the easiest and cheapest method. Stone is heavy; the shorter the distance moved, the fewer the hassels. Every extra mile traveled to obtain supplies means more truck wear, additional gas consumed and extra time spent.

Fortunate house builders find enough stone on their land but this is frequently not the case. There are a number of reasons why it may be impractical or even impossible to build with stone from one's property. For instance, wooded land crowded with undergrowth makes gathering tedious. It is difficult to walk let alone maneuver a wheelbarrow or motor vehicle through dense woods. If the stone is sparce then it may be hardly worth the effort to collect. Some builders have plenty of stone on their land but it is small or irregularly shaped. In a case like this, owner-builders must decide whether it is preferable to use inferior material from their land or a better grade of stone found elsewhere.

If there is little stone on one's land or if it is unsuitable for use, there are often other sources which may be explored. For instance, a nearby quarry may be a convenient source of cut stone. Although quarries generally sell their stone its price need not be prohibitive. A large part of the cost is in hauling it. If one goes directly to the quarry with a truck, not only does the price per ton decrease sharply but the stone may be selected personally.

Even when one is not gathering stone from a quarry it is a fascinating place to visit. Its scale is overpowering. Quarries often remove material from cliffs hundreds of feet high or they excavate far into the earth. These mines are testimony to people's ability to manipulate while they often destroy the environment for their own purposes.

One advantage of cut stone is its beauty. Quarried stone is usually cut or blasted from a monolithic formation of rock, freshly exposing each piece. Neither sunlight, rain nor air have dulled the colors or smoothed the rough edges. Each piece of quarried stone is being exposed for the first time since it became rock millions of years ago. This material cannot be surpassed for its brilliance or its exquisite patterns and textures. Also, quarried stone is usually excellent for building purposes. A typical piece breaks cleanly and forms flat surfaces with square or angular corners. Such material can be laid with ease and speed, offsetting its additional cost.

Stone is quarried for a number of purposes. Some quarries extract stone not for building but for crushing into various sizes - ranging from half-inch gravel to pieces, called rip-rap, weighing from twenty-five to two hundred pounds. Gravel is used for road covering or for making concrete while rip-rap is generally used to stabilize embankments. Although it is not quarried for use in building, rip-rap can sometimes serve this purpose.

If a structure is to be faced with stone there are specific quarries which produce veneer material. Veneer stone is mined in large sheets not more than four inches thick. Later in this book it will be explained how these shapes are laid on edge. This method of stonework must not be used structurally for it is too thin, but it is effctive as a decorative surface. If one plans to build a monolithic stone wall veneer material should not be purchased. To build a monolithic stone wall you will want stone from a quarry that extracts pieces which are eight to twelve inches thick. Often the same quarry will produce both veneer and structural stone.

In spite of the advantages of quarried stone some owner-builders prefer to use fieldstone. Fieldstone has a different character. The shape of quarried stone is artificially created while that of fieldstone occurs naturally, its shape changing through continual exposure to the elements and life forces. The cut surface of quarried stone has a fresh brilliance while the surface of fieldstone appears time-worn and venerable. Fieldstone has an aspect of integrity and naturalness which contrasts that of cut stone.

Fieldstone is found in many shapes and sizes, colors and textures. Often covered with dirt the character is indeterminate until the stone is washed off. Notice how it is embedded in the ground, how firmly it rests there; and imagine how long it had been in that place. Rocks are passive and if their beds are stable they remain in place. Remember this as you lay stone in your own wall.

Farmers view fieldstone in a different light from that seen by builders. Most farmers will encourage you to haul stone from their fields and pastures - an excellent source of this material. To them, stones are merely objects which have been breaking their plows and straining their backs for generations. Often the landowners and their ancestors have already done most of the work by piling stone along the borders of plowed fields or around trees in pastures. One has only to carry it from these piles to a truck.

As stone is removed from these piles remember that people are not the only inhabitants of stone shelters. Stone piles often house creatures ranging from mice and ants to blackwidow spiders and poisonous snakes. Whenever you are collecting stone it is wise to wear gloves and watch where you step.

Another source of fieldstone is found along roadsides. The shoulders of old dirt roads are often lined with stones graded to the side during years of maintenance. Two people, one walking alongside and one driving, can quickly load a truck. But do not overload the truck for stone is heavy. What may seem like a small quantity can easily weigh over a ton. When loading, constantly check the truck's springs. It is tempting to throw on just one more piece - which could break the axle.

Abandoned mining operations are another good source for locating building stone. These vacated works are shown on maps available from the United States Geological Service. Such stone has already been reduced to useful size and tends to be piled, making it handy for loading. You will probably have to pick through these piles, referred to as tailings, for they will contain quantities of unuseable debris.

Sometimes one happens to find an old foundation or a chimney of stone. What a prize! Not only is the material stacked in one place but each stone has already proven useful. Most old foundations or chimneys are built dry (without mortar) or merely with mud, so dissembling them is not difficult. They have only to be taken down, course by course.

Getting permission to remove these old structures is likely to be the hardest part of the whole process. Often an old chimney will be all that remains of an ancestrial home and the present owner may be quite attached to this monument from the past. More than once a prospective stone collector has found that a coveted old drywall marks the boundaries of a still consecrated family cemetery.

If permission to dissemble one of these structures is obtained, much can be learned about masonry while you take it down. As each stone is removed, pay close attention to how it was bedded in relation to other stones around it. The mere fact that this structure is still standing shows it was well built. Or if it is in partial ruin perhaps the reason will be indicated through your observation. In any event, these structures should be taken down with reverence for a bit of history is being erased.

Besides quarried, mined and fieldstone, there is another type of stone available in most areas. River and creek beds, flowing or dry, often have a quantity of easily accessible stone. Like fieldstone, river stone is a product of its environment. The surface of stone found in creek beds is worn smooth by water which has washed over, under and around it. This polished surface often reveals beautiful patterns and colors within. Sizes and shapes tend to be as varied as fieldstone. One may find both large, square shapes and rounded cobblestones. An additional advantage of creek stone is that it is often clean and ready to use when removed from its watery bed.

As experience is acquired in stone masonry one develops a knack for collecting the right pieces for the work at hand. To a skilled mason gathering stone is like choosing parts of a puzzle. Even as stone is thrown into a truck, the mason anticipates how various pieces may fit into the work. Much effort may be saved by being selective when collecting this material for not every stone will suit a mason's building style. Indeed, the kind of stone that is found will influence the system of construction to be used. When facing a wall thin flat stone is used. A solid wall requires bulkier, squarer stuff while round stone is better used in a formed wall. And one may encounter stones so irregular that they are only useable as fill.

The following chapter will discuss in detail how to study various pieces to determine their usefullness. As you collect stone, remember that all sizes will be needed. Those small, seemingly insignificant hand-sized ones will be important for filling gaps between large, beautiful boulders - those which may require two people to maneuver into a truck or into place during construction. Ultimately, there is a use for just about every stone you gather.

creek stone

field stone

quarried stone

cobble stone

Looking at Stone

To build with stone you must learn to look at each piece in a new way. It is important to become thoroughly acquainted with every stone you intend to use. Pick up one and turn it over in your hands to see and feel its entire surface. One of the qualities you will observe is its color. Some stones have a variety of colors. They will often form a regular pattern indicating veins of quartz or mica. Such veins could reveal faults along which the stone may break. Some colors are merely stains on the surface while others reveal the stone's composition. For instance, reddish stone often contains iron while that with a greenish hue could bear copper.

The texture, weight and shape of a stone are also important. In some there is a grain, much like that in wood, but others have a sandy texture which crumbles easily when handled. It might have a smooth polished surface or it could be rough or jagged. These qualities will contribute to the appearance of the finished wall. All stone is heavy but some is more dense than others. Its weight may hint of its overall usefulness as does its size and shape. Large or small, round or flat, angular or worn, a mason pays close attention to these features when choosing stone.

In addition, stone is viewed in terms of its function. Each piece must relate to others around it. Any one set in a wall is doing at least three jobs. It sits solidly on stones placed below it. It provides a firm bed of support for those above it. and it presents an outer face which gives the wall an attractive appearance.

A mason must evaluate three aspects of every stone to determine how it will be used. A stone must be judged for its base, top and face Illustrated here are some examples of these essential features of building stone. Every stone must have a solid base which rests firmly on the bed provided by stones below it. Each must also have a flat top surface, providing firm bedding for stone resting upon it. The base and top are called its building surfaces. At right angles to both building surfaces is the face. It is the only surface visible once the stone has been laid. When selection is made it is tempting to choose the attractive face but remember that, structurally, the building surfaces of the stone are more important and therefore of first priority. Don't let your head be turned by a pretty face with nothing behind it.

Stone may be used for building in many ways. All those shown above could be turned around so that a different surface might be assigned to function as base, top and face. An example of this would be a stone which, when laid flat, could be used in a solid construction but, when placed on edge, becomes excellent as veneer.

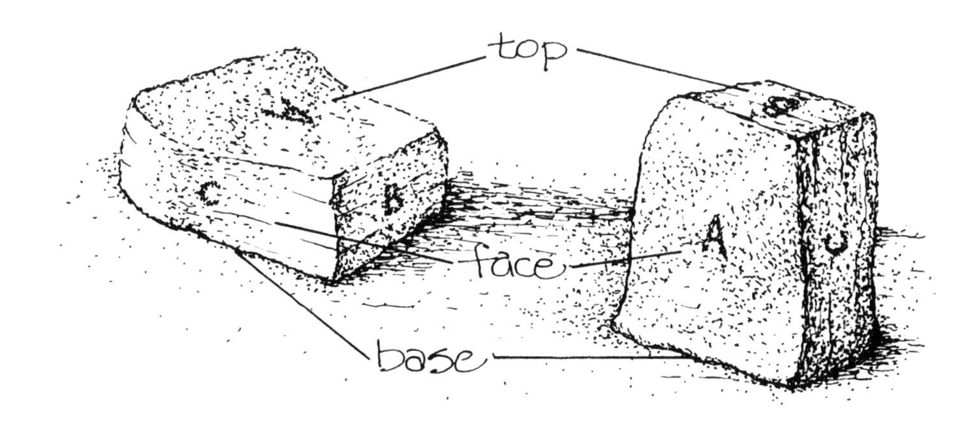

the same stone turned two ways

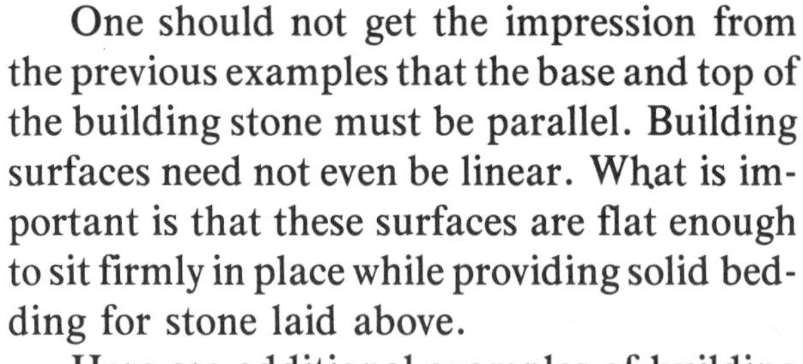

One should not get the impression from the previous examples that the base and top of the building stone must be parallel. Building surfaces need not even be linear. What is important is that these surfaces are flat enough to sit firmly in place while providing solid bedding for stone laid above.

Here are additional examples of building stones whose bases and tops are not as well defined in terms of squareness as are previous examples. Stone A has a linear base and top although the planes of these two surfaces are not parallel. B has a linear base and its top, though curved, is still reasonably flat. When this piece is placed in a construction and surrounded by other stone there will be no problem to build on it. Stone C has a flat, linear top but a curved base. This piece can be laid securely if it has a correspondingly concave bedding from which it may never move.

Stone D has a flat base, little top surface and its face is triangular. This does not mean it is poor building material. The sides of this triangular-shaped piece are flat enough to support any stone leaning against it. This same piece could be flipped 180 degrees to use its flat top and wedge-shaped base.

Being circular, there is no distinct place on stone E where the base stops and the top begins. Still it is excellent building stone. If properly arranged it will seat securely and support other stone. This illustrates the point that it is not angularity but the quality of

A

B

C

D

E

building surfaces that makes good building stone. If its top and base are reasonably flat and at right angles to its face then the stone will function effectively in a wall.

Generally it will not be as easy to identify a stone's functional parts as those we have examined thus far. Most do not have such well defined surfaces. It often takes imagination to anticipate how a stone can be used for building.

Can you see the base, top and face of each of the stones pictured above? To develop a mason's eye requires viewing complex shapes as basic, geometric forms. Squinting will help you to see the stones in terms of their definitive parts. An experienced mason might see them as illustrated below.

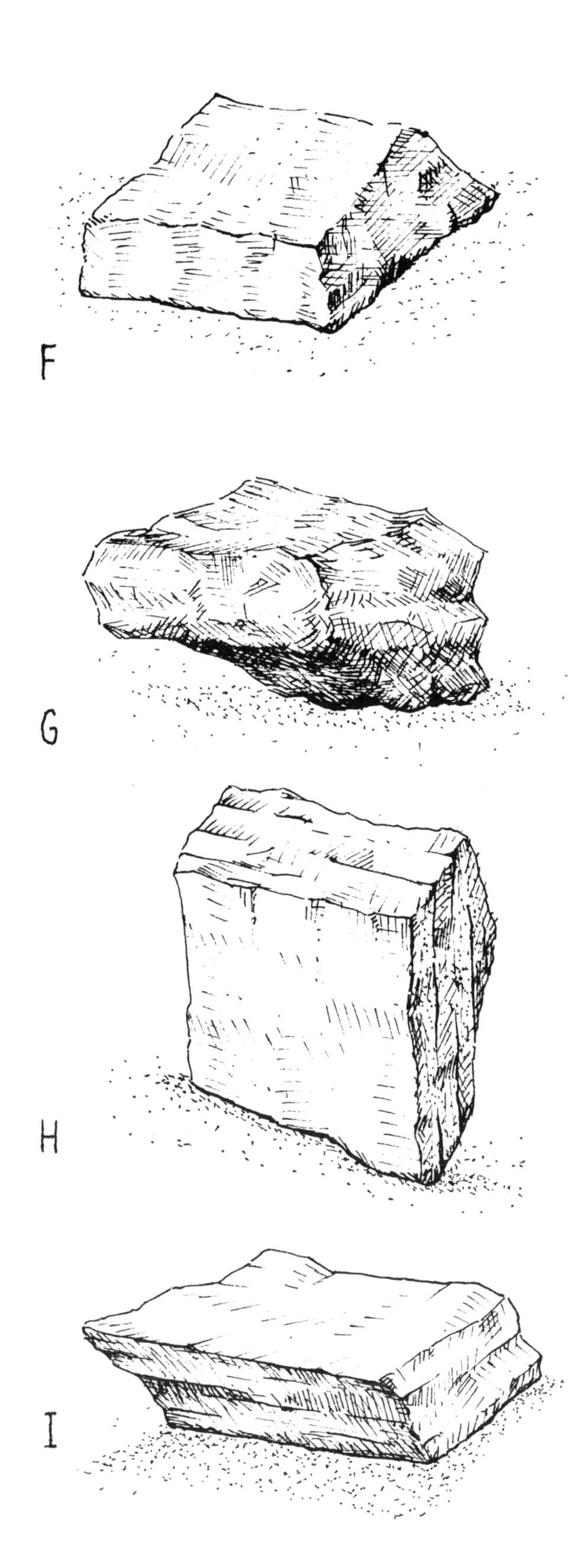

These stones do not have the clearly defined parts of the previous examples. One surface tends to melt into another but the parts are still there. One has only to learn to detect them. This is not to say that one can find every kind of desirable building surface in each stone. If fact, the commonest problem for beginning masons stems from their attempts to build with unsuitable pieces. Some stones simply cannot be used for building. A mason must, therefore, learn discretion. Rejection of some irregular stones should be obvious. Other stones deceptively appear to have useable building surfaces yet will cause problems if used. Those in the next illustration seem, at first glance, to have the requisite surfaces, but nevertheless they are unsuitable for use in building.

Stone F seems to have flat, substantial building surfaces but its top is not perpendicular to its face. This makes F structurally unacceptable. Any stone placed upon this piece will tend to slide forward. Stone G, on the other hand, has a flat top perpendicular to its face, but the base is so irregular that it would need a very unusual bed on which to seat firmly. H is veneer stuff which would be difficult to secure in place because its base is not at right angles to its face. Even if one did manage to seat it in a wall, it would be virtually impossible to place stone above it because the top surface slopes forward. Although stone I will remain in place, its face is perpendicular to neither its top nor its base. This means the face of this stone will not sit plumb with the surface of the wall. This is not a structural detriment but the builder may reject this piece for aesthetic reasons.

Some stones, like those shown, can be altered for use in building. They may have to be turned to a different side or be reshaped with a hammer and chisel. Shaping will be discussed later.

Various uses of shims

Another way to make unsuitable stone useful is by sliding a shim beneath it, leveling the top and plumbing its face. Obviously, shims are structurally integral to the larger stones they support. When selecting pieces for shims make sure they are hard and unlikely to crumble under the weight of the finished wall. Shims can be employed in a variety of ways, as shown below. The act of shimming a stone is opposite to that of chipping a stone. Instead of removing part of a stone to create a useable shape one adds to it, making a useful whole from two pieces.

There are times when a mason must find more than just three functional areas on a stone. One such occasion is when cornerstone is selected. In most projects the structure will have squared corners. The skill and neatness with which corners are constructed is important to the overall apperance of one's work. Corners should be carefully laid because they are the most vulnerable part of a wall.

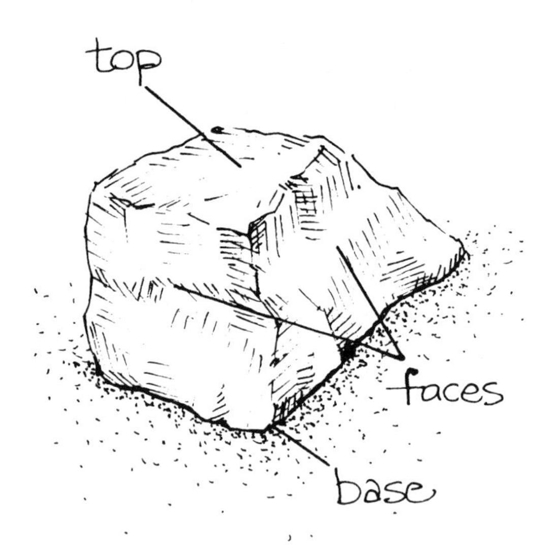

A corner stone

At a corner two sides of a stone are exposed, therefore a cornerstone must have two presentable faces with a flat base and top perpendicular to each face. Furthermore, the two faces must come together at approximately a right angle. Sometimes, however, one may build acute or obtuse corners requiring cornerstones with a corresponding angle. This specificity makes cornerstones hard to find. As choices narrow one spends increased time looking for a particular piece. To build a small-based structure like a chimney, it is not unusual for a mason to spend half the building time constructing corners.

At this point the reader may be wondering if the back and sides of a stone are important. Generally, when stone is laid in a wall, the sides and back do not support weight and are not seen. To this extent they are unimportant. But there are specific cases when the sides of a stone must be selected as carefully as its face, top and base. One example of this occurs when building an arch. Here sides do support weight and must therefore be as flat and perpendicular to the face as the top and base surfaces are. Stone which meets such requirements is hard to find in its natural state. One usually has to shape it with a hammer and chisel to suit the need.

The art of seeing stone is by far the most elementary yet the most difficult part of learning stone masonry. It is developed only with time and experience. As you learn to properly select this material working with it follows naturally.

Two Rules

Each of the stone walls pictured on these pages was built in a different manner. The dry wall is composed only of stone - no mortar - and there is an example of mortared stone packed in a form. One of the walls is a four-inch veneer covering cement block and another was laid with a solid twelve inches of stone and mortar. Whichever method of construction is used there are two basic structural rules which must be followed. These will be discussed in this chapter. The best way to learn them is to first understand the nature of their origins.

The most important factor determining whether a wall will ultimately stay in place or fall is not the quality of the mortar or the hardness of the stone; not the tightness of the joints or the firmness of the footing. These variables are all of consequence, but the principle force which determines whether a wall stands or falls is *gravity.* It is a simple fact that if you defy gravity you will eventually lose the contest.

Discussion about gravity seems fundamental and, in fact, it is. But, nevertheless, the beginning mason will often try to stack stone in a manner opposing this all-pervasive force. Stone is heavy and falls earthward due to the pull of gravity. The objective of wall building is to set stones in such a way that gravity will hold them in place; to arrange them so that they are, in effect, falling on one another.

To illustrate this point we will examine some simple piers, like those supporting the corner of a house. Both stone piers shown here are now standing but one will endure while the other will eventually fall.

Gravity exerts a constant force on the stone in these piers. The structure on the left uses gravity to its advantage. Each stone falls straight down, weighting the one below it and stabilizing the entire structure by holding each piece stoutly in place. It has been carefully chosen for its flat, firm building surfaces. When a house is built on piers having this kind of organization, the weight of the building insures that none of these stones will ever move.

The right-hand pier, to the contrary, works against gravity on every course. Observe the arrows illustrating the direction in which gravity pulls at each stone. Each tends to slide off the one below it. As additional stone is stacked the tendency to slide becomes greater. Friction is all that holds this pier together. In time, as more weight is added the force of friction may be overcome. The weight of a house could cause the pier to slide apart.

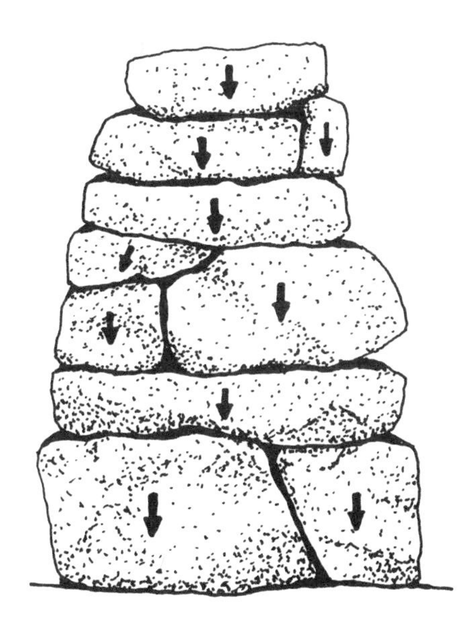

These two piers illustrate the *first rule* of stone masonry; always provide a firm, flat bed for each stone to be laid in place. It could be said that masonry is the process of preparing a bed for the next stone to be laid. The illustrations used here are examples of dry wall but these rules apply to all methods of masonry construction.

A wall may be thought of as an elongated pier. The principles of construction for the two are the same. The cross-sectional diagrams of the two walls illustrated here should make this clear. The wall on the left is well built. Each stone is placed so that it either falls solidly on the one below it or the force of gravity pushes it inward. None has a chance to slide from place. Notice that, unlike a pier, the larger size of the wall permits a greater variety of shapes. Not all the building surfaces on these stones need to be parallel or even linear. Varying shapes work together for stability

Although the face of the wall on the right looks cohesive and stable, a cross-sectional view reveals it is poorly built. The stone's bedding is not level and gravity will tend to pull most of them outward and downward. If this wall were disturbed in a number of ways (either by additional weight, vibration or the freezing and thawing of rain water in the joints) it would slide apart in time.

This basic rule of stone masonry is formulated by the way stones naturally compose themselves. To test this idea, try an experiment as you unload stone from a truck. As it is tossed to the ground observe how the stone lands. The first pieces thrown will hit the ground and roll until they come to rest on their natural base. Before long the ground will be covered with pieces snuggled against one another. The pile then grows as more stones are cast onto it. They hit, bounce and finally settle on their natural bed. By the time your truck is unloaded, tossed stones will have created a stable pyramid. If this naturally-formed pile is examined much will be learned about building with stone.

Now examine two more piers. Both appear to be skillfully stacked conforming to the first rule of masonry. Yet the pier to the right is stronger and more enduring than that on the left. Each of the stones in the right-hand pier is linked together, working as a unit. The weight of each piece is distributed over at least two others below it. Each supports the weight of at least two others above. The left-hand pier is actually two separate piers nudged closely against one another. They are two independent systems, both tall and spindly. The left-hand pier would have been much stronger if its two elements had been unified.

These two piers illustrate the *second rule* of stone masonry which requires that stone be placed so they distribute their weight over at least two others below. This is the principle of crossing joints to insure that the strength of a masonry structure is maximized. Each stone is secured to all others. This principle is apparent in the construction of a simple structure like a pier, but one must be more observant when building a wall - especially when all stone is not square.

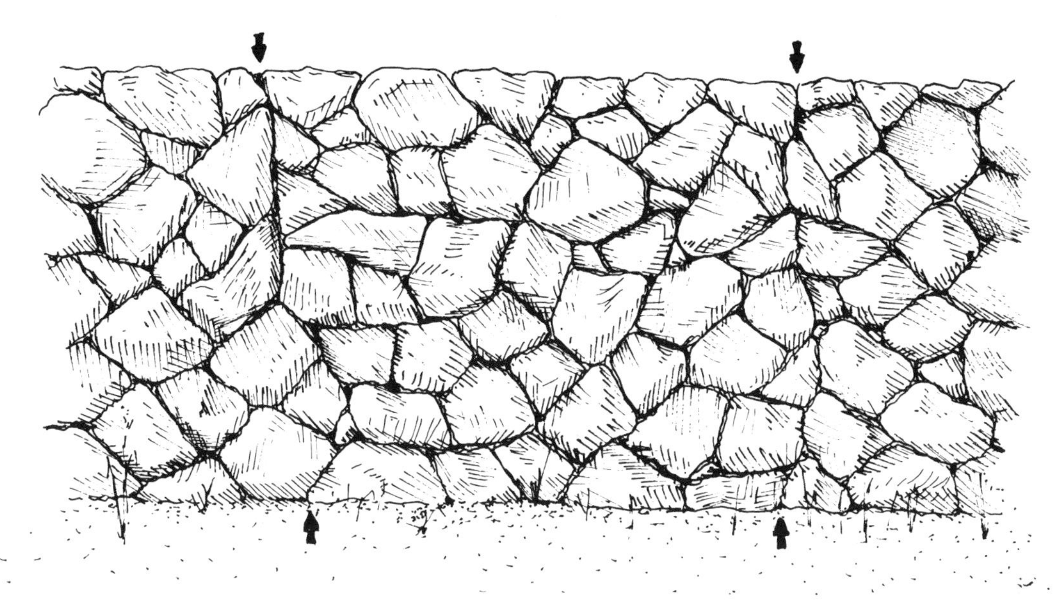

Notice how the mason who laid this wall inadvertantly built three separate walls. At first glance it seems to be a unified structure, but arrows at the bottom and the top of the wall reveal lines of uncrossed vertical joints. The wall in effect is constructed in three sections and, as a result, is considerably weakened. If this were a retaining wall the arrows would indicate the points at which the wall would first give way to outward pressure from behind.

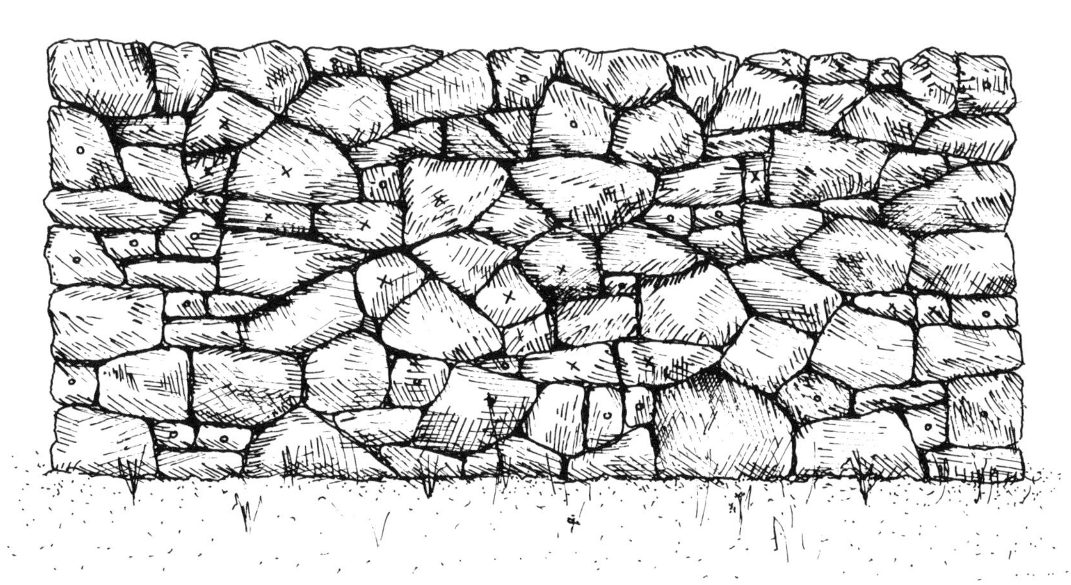

Examination of this wall reveals that not all stones distribute their weight over two others. In cases such as those stones marked o, no substantial strength is lost. These are not errors in workmanship but conscious exceptions in observance of style. In other instances where stones are marked x, strength is lost through failure to cross a joint. Here, larger pieces could have been chosen so that their weight would have been more equitably distributed.

These are the two basic rules for creating stone masonry. To understand them is to know what makes a viable stone construction. If they seem elementary, they are. The theory is simple but application is another matter.

If you look around you will see that these rules are ignored time-and-again, and when putting up a masonry structure you will realize why these rules are so often disregarded. For instance, when building a wall you may find a stone which neatly fits the space to be filled, although it has a flaw. Its top surface slopes downward, toward the outside of the wall. It is tempting to use it nonetheless since, for some time, you have been trying to fill the space. But beware, more time will be taken trying to solve the new problem it creates for you. And no solution using this particular stone will ever be stable. Occasionally a stone is not quite long enough to adequately cross the joint below it, but one is tempted to use it anyway. Do not.

The consequences of ignoring these rules are often not immediately apparent. Unstable walls may stand for years before gravity topples them. The introduction of portland cement for use in masonry construction has made it possible to defy gravity longer than previously. However, if one looks around with a critical eye, numerous examples of crumbling walls will be found, testimony to the fact that basic rules were not followed. For a lasting, maintenance-free structure - as stone work should be - it is best not to try to thwart gravity but to work with it.

Fitting Stone

The two rules discussed in the previous chapter give all the information you need to build a strong, durable stone wall but there is another aspect of stone masonry not yet covered. A major part of the appeal and mystique surrounding stone masonry is not necessarily due to the creation of an enduring structure but rather how its stones fit together. Of course, to some extent form does follow function. For a structure to be strong its pieces must fit snugly, but the varied shapes and sizes of stone permit literally countless ways to fit it together. Every mason has a particular style and, insofar as structural rules are followed, no one style is better than another.

The photos on these pages illustrate different styles of stonework. Some masons prefer square or geometrically shaped pieces while others work with more fluid, organic forms. There are those who prefer to build with large stones while others use a variety of sizes. Many masons organize stone in structured courses while many casually lay random patterns. To some an attractive wall contains stonework with carefully defined joints. To others rougher construction seems more in keeping with the nature of stone. Some like flat surfaces; others purposely build walls with jagged features. Variations are countless.

Discussion in this chapter will not concern any particular style but will describe how to fit stones so that you may determine your own. As in the previous chapters, suggestions given here apply to all methods of stone construction—laid, faced and formed.

When fitting a stone one looks principally at the shape of its face. This does not mean that its base and top may be ignored. To be sure, even before considering the face it must be determined that the piece has both a solid base and a reasonably flat top. This is always the primary concern but, once this is concluded, one turns to the stone's face. This is the visible surface of building stone, the only surface exposed to rain, ice, wind and the scrutiny of those who observe your work. How the wall looks is mainly determined by the way in which the stone faces fit against one another. In essence fitting stone means arranging their faces to one's liking with the limitations of the structural rules.

Building a stone structure is by necessity a sequential process starting with the footing upon which the first course is laid. On this course other pieces are placed, one at a time. This process of laying stone upon stone is continued until the desired height is reached. On the last course one usually finishes or caps the wall by selecting flat, level pieces.

When watching a mason at work it may seem as if one stone at a time is being laid. However, the mason is always thinking about more than the stone being handled at the moment. Structurally and aesthetically, a wall is a unit resulting from individual pieces working together. When you lay a particular stone, you must ask yourself a number of question concerning the entire structure. These are:

1. Does this stone sit securely on its bed and does it unify the stones below it?

2. Does the top of the stone line up with its neighbors' tops so that all pieces work together to form a sound bed for stones that will be placed on them?

3. Does the stone fit well in the space to be filled?

4. Is the size, shape and color suitable to the effect one is trying to compose?

All this may seem like a lot to keep in mind but, as one gets into the rhythm of stonework, these questions are asked and answered more or less automatically. For now we will explore them, one by one.

The first question should be familiar. It basically asks whether one has adhered to the two structural rules of the previous chapter: does the stone sit solidly on its bed and is it distributing its weight over two stones below? To consider this question, lay the selected piece in the spot to be filled. Does it sit firmly

in place or does it want to roll off? Perhaps it sits relatively well but tends to rock unsteadily. In that case, one must decide whether or not it can be stabilized with a shim or with mortar. The face of the stone does not directly affect how it will seat, but the shape of the face will give an indication of its potential fit. For instance, if the bed is concave, look for a stone with a convex base and a face that is curved on the bottom.

The second part of this question may be answered at a glance. If the base of the selected stone spans a fair portion of the two below it, then it will tie them together. If it does not unite the two below it, the stone may be detrimental to the strength of the wall. Clearly, the answers to this first question are not a qualitive yes or no. The process of fitting stone is one of compromise. Rarely do stones occur in just the right shape. One's search for the best possible choice must be within the limitations of the supply, the available time and one's personal tolerances. No matter which stone is selected there will always be another that would fit as well—or a bit better.

The second question of the series asks whether one has planned ahead for the next course to be laid. This is important. If a mason anticipates in this manner a situation will never occur where a space, impossible to fill, results from prior thoughtless work. Building flat beds saves much time for the mason does not have to search for odd-shaped stone to fit irregular beds. To clarify this point, suppose you are working on a section of wall resembling this illustration.

You are searching for a stone to fill space A. The bed for this space is well prepared. Any number of stones exist that have a curved base which will sit well on this concave bed. Stone B is selected and set in place. It fits well on its bed but what will happen when you build on it? Obviously, it would be difficult to span B and its neighbor to the right. There must be a better choice. C is slightly more desirable. One could span it and its neighbor

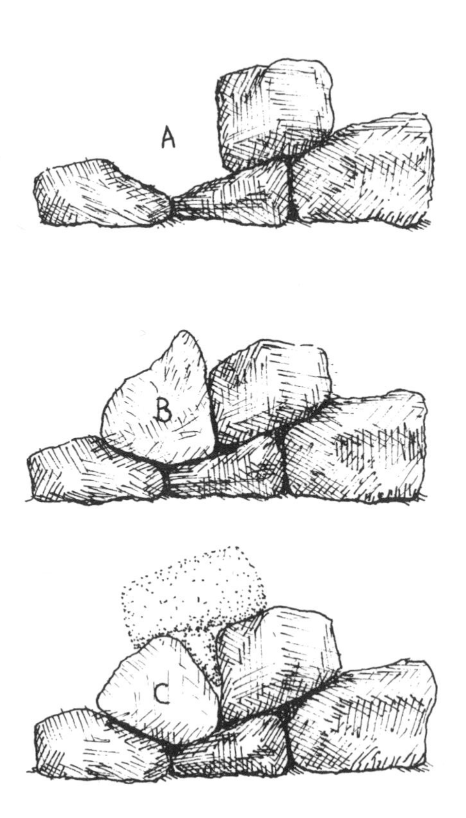

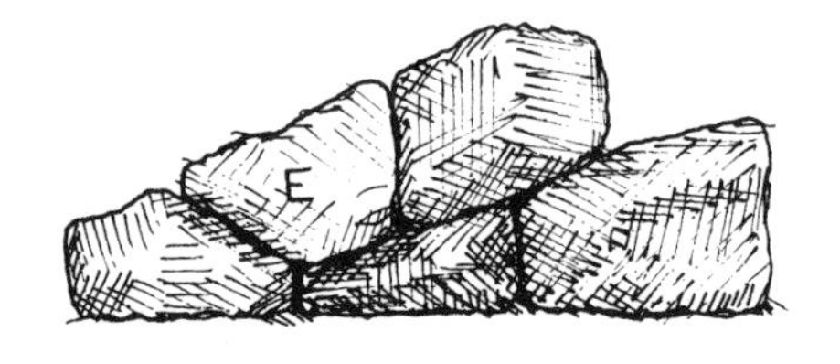

with a small triangular filler and one longish stone having a slightly concave base. But experience has taught that concave stones are not easy to find so the search continues. If a better stone cannot be found one can beg the question with a piece like D. Stone D at least creates the sitution whereby one merely needs to find another small stone to place on top of it. Or persistance may locate E or F. Either choice prepares sound bedding for the next course. Stone E sets up a bed for a piece with a linear base, while F creates with its neighbor a cradle-shaped bed in which to lay a stone with a convex base. By now it goes without saying that the tops of these stones must be flat-surfaced.

As you gain experience you will begin to notice that in any pile some shapes are more prevalent than others. Some stone is predominantly flat and square. Other may be roundish, with convex surfaces. Very little stone will

have concave surfaces. It is wise to check through a pile before starting to build, noting the predominant shapes. After that, one can arrange to build spaces which use these shapes, avoiding building spaces which demand hard to find shapes. Building will thereby be faster and easier.

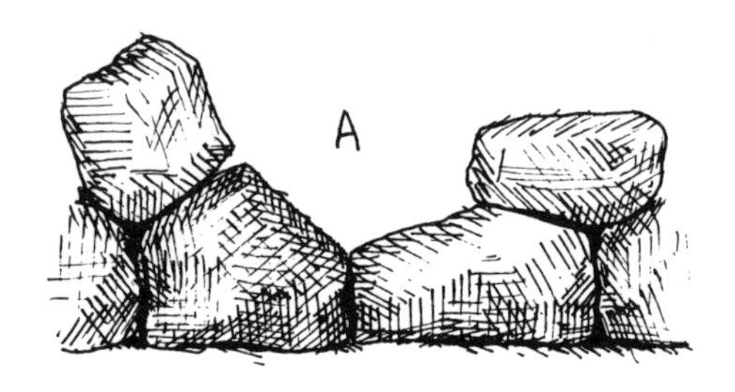

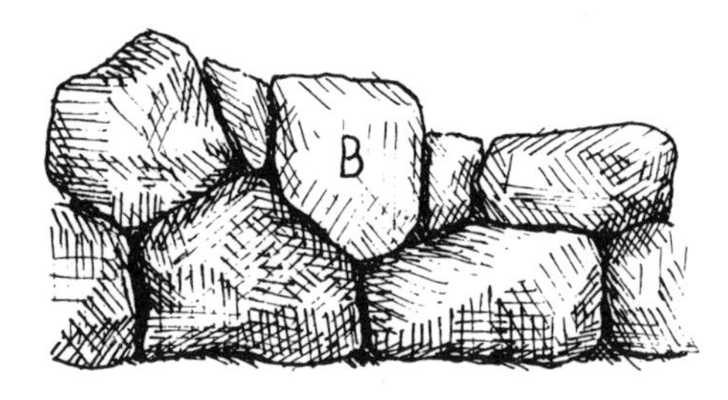

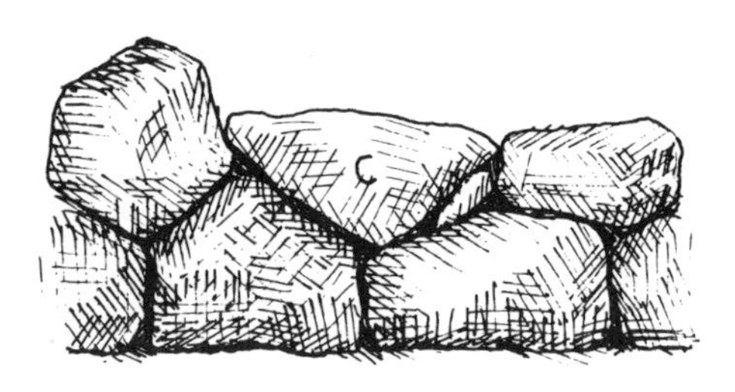

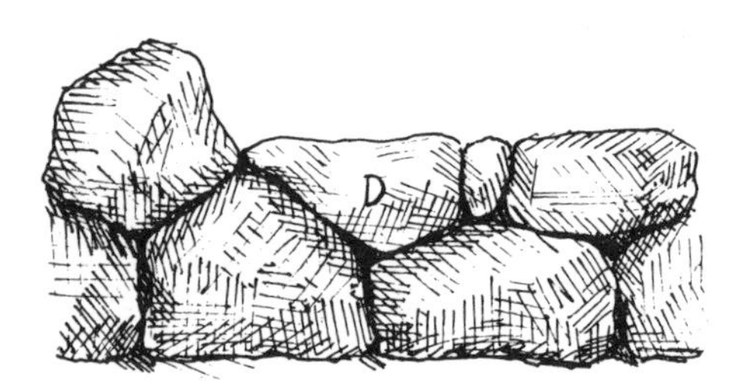

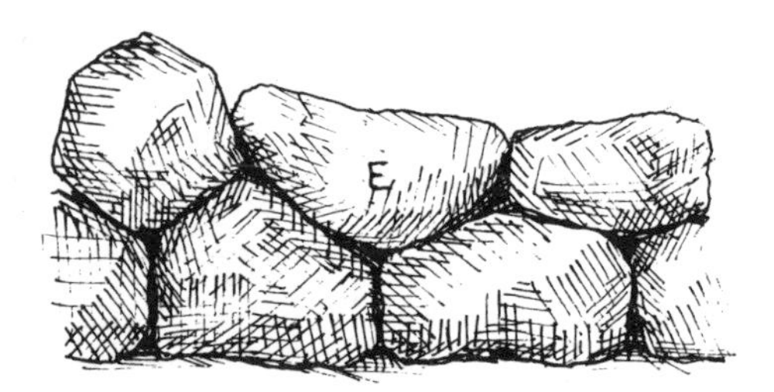

The third question is an aesthetic extension of question two. It asks whether the face of the stone to be laid looks good where it is placed. The answer is entirely personal but we can discuss the process by which your answer may be reached.

Take another specific situation in which a section of a wall under construction resembles this drawing. A stone to fill space A must be found. How do you look for this particular shape? Fix in your mind a geometric figure approximating its form. For space A one might picture a piece having a face with a V-shaped or rounded base and sides that angle out. The top edge of the face must line up with one of its neighboring stones but its shape need not be clearly defined. In other words, envision a stone with a semi-circular face.

Next, go to the pile to look for stones with faces having the simple shape you envision. Your first option may resemble stone B. It seats firmly and its top lines up well with its neighbor's to the left, but there are large spaces on either side between it and its neighbors. To be structurally sound these spaces would have to be filled with small, thin stones. The decision to use B will be an aesthetic one. Suppose B is rejected and stone C is located. It fits snuggly against its neighbors but leaves a gap beneath. This gap could be filled with a triangular chinker and the wall would remain strong. A better choice might be to use stone D in combination with a filler piece or stone E. Some masons may find both these solutions unacceptable. Stone D would be rejected because it calls for a filler stone, and E would be refused because it is basically round and leaves large, triangular gaps at the joints. The choice is the individual mason's.

Perhaps you are feeling at this point that there ought to be a surer way to select individual stones rather than by stumbling around a pile, randomly searching. There isn't. All one can do to make the job easier is to sort stone for its general size and shape. Then at

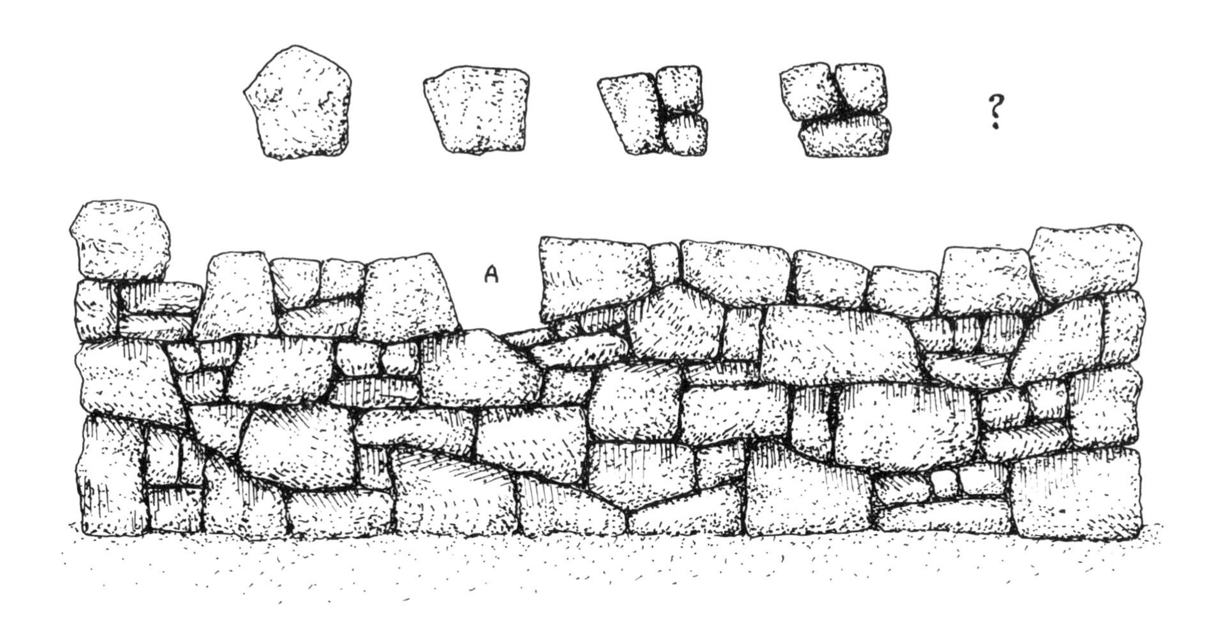

least it will not be necessary to scrutinize each stone every time. Take comfort in the fact that as your stonemason's eye gets keener, the time spent searching for a particular shape gets shorter.

When first starting to build with stone it will probably tax your patience to find one stone to fit a particular space. But as experience is gained you will want each piece you lay to fit the overall pattern of the work. At this point the final question is asked and answered and your style develops.

To clarify the question we will probe another example. Here is a partially built wall in which the mason has set up a regular pattern using a distinctive style. Whether the reader prefers this style or not is irrelevant. The next move is to fill space A and there are a number of ways to do it. A few alternatives are shown. Which of these four do you think best fits the style so far determined? Which alternative would you select if you were building this wall? There will, of course, be no precise answer to this question. The purpose of asking it is only to show that there is more to laying stone than merely finding a piece which fits the space.

Stone masonry can be as sophisticated or as simple as one cares to make it. Some masons choose not to plan ahead at all. They just let the stones fall where they may, allowing a natural pattern to emerge. Others get satisfaction from creating well thought-out designs in which the shape of every stone laid is a conscious decision.

↑ good   bad →

There is one aspect of stone masonry requiring particularly careful forethought: the finishing or capping of the top of a wall. Generally, the goal is to make the top of a wall level or reasonably flat. There are, however, numerous exceptions to this goal. As the top of a wall is approached, the mason should think about capstones and prepare for their positioning ahead of time. Too often the style of an otherwise well built, handsome wall is disrupted because arrival at the top seems to have taken the mason by surprise. The main body of the work may be made of fairly large, angular stone but then, abruptly, at the top of a wall one finds a row of thin, flat pieces obviously placed there only as an afterthought to level the wall. How much better it would have been if the mason had thought ahead and taken the time to use the same sizes, shapes and style throughout the work. A wall should end effortlessly. Masons generally agree that stonework can be judged by how it is capped.

The art of fitting stone is the great intangible of stone masonry. It is that aspect which differentiates one mason from another. The suggestions given above will hopefully make it easier to get started in this work. But words and even pictures do not substitute for experience. The only way to learn to fit stone is to do it.

Shaping Stone

Until now discussion has been about stone in its natural form - the way you will find it when taken from a field or quarry. The previous chapter about fitting stone bore the assumption that somewhere in a pile of stone would be a suitable piece to fill any space. Suppose, however, that when building you cannot find a stone to fit a particular space. If you are attempting a specific design it is likely that this will occur. The choice then is between hunting more stone or shaping that on hand.

Many owner-builders are reluctant to try to shape stone. Some do not want to intrude on the natural forms of the material they are using. More often, though, beginning masons avoid this skill because it appears to be slow, tedious work. Shaping stone *is* slow and tedious, especially when one is first learning. It breaks the rhythm created by placing stone upon stone without interruption.

As experience is gained, however, shaping becomes an invaluable technique for the solution to problems of laying stone. Once adept with a stone hammer it can become routine for the mason to knock off rough edges from nearly every stone handled before putting it in place. In fact, shaping is often more efficient than wandering about a pile in search of suitable pieces which may not be there. Shaping stone also gives the mason a new element of control over his work.

To learn to trim stone it is necessary to become acquainted with it in a different way than previously experienced. Of course, you still pay close attention to shape but now the factors of texture, grain and hardness become major considerations.

Pick up a stone and take another look at it. Is it solid or does it have a crack running through it? Does it feel hard and dense or is it soft and sandy in consistency? Does it have grain, like wood, along which it is likely to split or is it homogeneous in character, like soap? These are qualities which must be noticed before one attempts to break stone. As will be quickly realized, one cannot work stone contrary to its natural inclination. One has to learn to break it in the direction it tends to split.

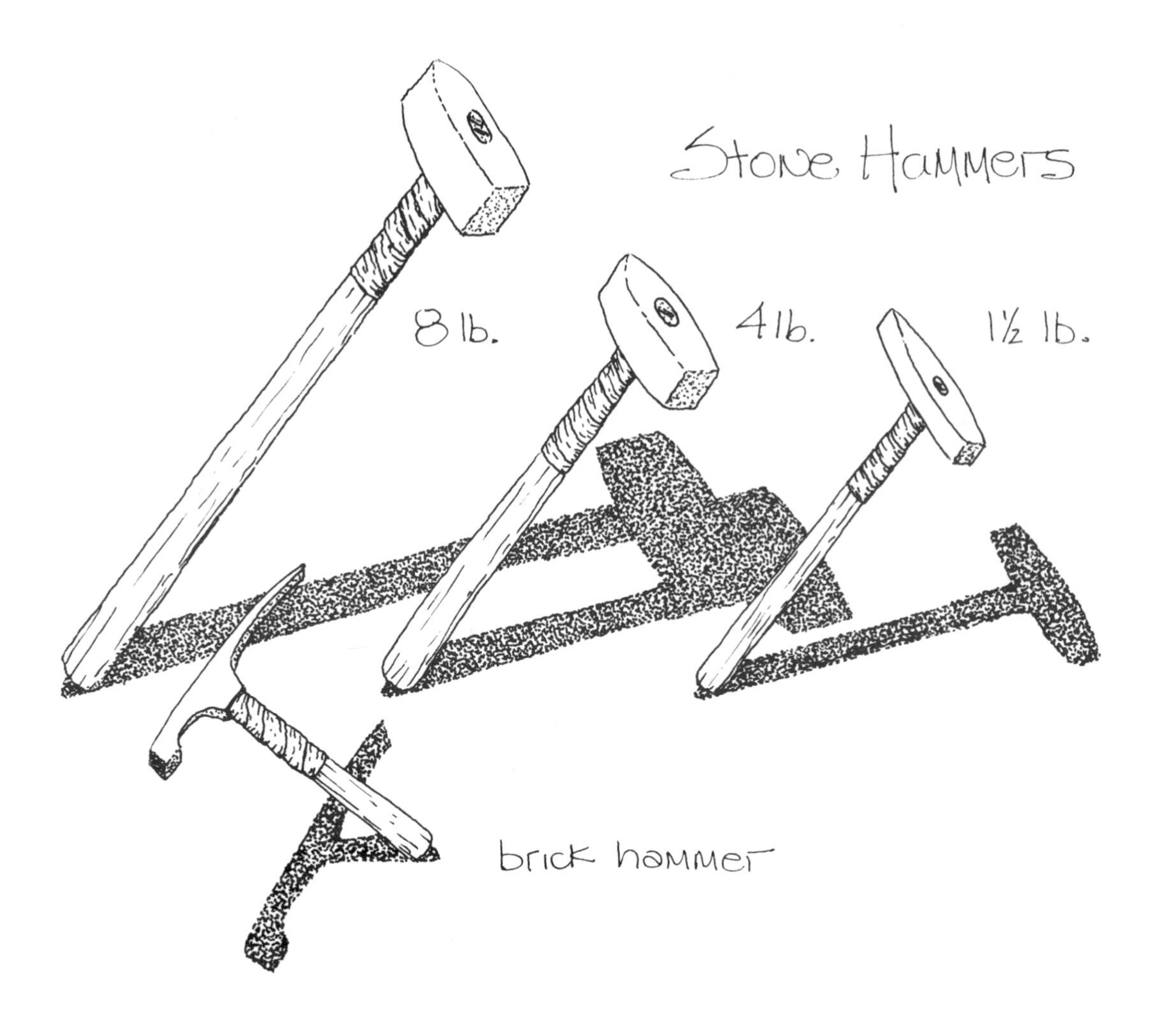

Quarries and large construction companies use special diamond-tipped saws and hydraulic splitters to shape building stone. The owner-builder will not have such tools and must use hammers and chisels for this work. The most commonly used tool for shaping stone is the stone hammer. These hammers come in three sizes and each has its purpose. The eight-pound, long handled sledge is used principally to roughly shape large stone or break it into smaller pieces. The four-pound hammer has a shorter handle, twelve to fourteen inches in length. This is the utility hammer which is used for chipping most stone. Finally there is the still smaller one-and-a-half-pound stone hammer. It is used for chipping and shaping small pieces or for trimming the edges of flat stone.

These hammers are each shaped similarly. They have a flat face which is square-edged and the head tapers to a wedge-shape. It is a good idea to wrap friction tape or, better yet, wire around the handle where it joins the head. This will protect it if you miss and strike the stone with the handle. Stone batters metal but it destroys wood.

Most stone chipping is done with the square edge of the hammer's flat face. Notice that the term "chipping", not breaking or splitting, is used. Stone shaping is necessarily a gradual process. Rarely does one break stone into a prescribed shape with one clean, sharp blow. Usually, struck in this manner it breaks in the wrong place.

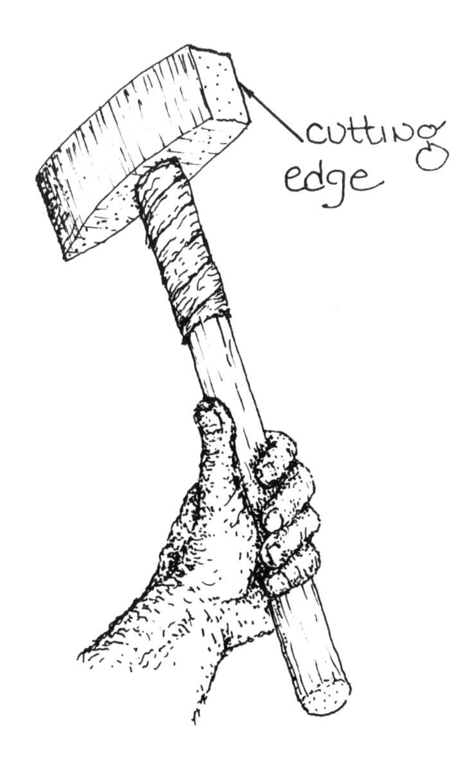

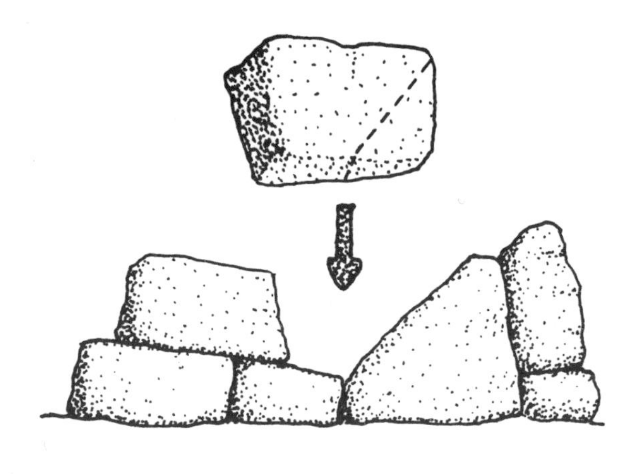

The correct method for shaping stone is to gradually knock chips off its edge until it is whittled down to a desired form. Suppose a wall is being built and a space is to be filled, but no stone on hand seems to fit. One piece, however, comes close to the desired shape if one corner can be removed. Before you chip this stone examine it. Is it solid? Determine if there are cracks in strategic places. Cracks mark what will be the initial breaking point. Does the stone exhibit a grain? The job will be easy if it breaks along the grain but troublesome if one tries to work against the grain. Finally, try to judge how much work it will be to remove this corner. If the stone is composed of granite it may be necessary to hammer for some time before it yields. If it is sandstone the corner should easily break. With this particular stone the corner is fairly thin. Using the four-pound hammer the mason's effort should not be extreme or shattering may result.

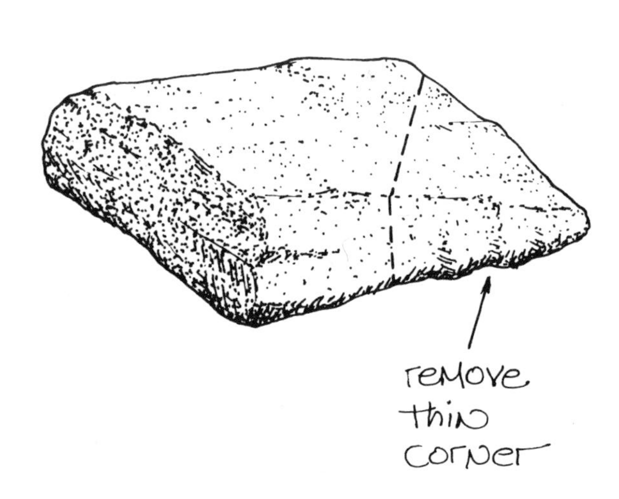

The secret of successful chipping is patience. Do not start by hitting the stone at the place you want it to be severed. If this is done it may break elsewhere. Place the stone in a bed of sand, on your knee or hold it in your hand if it is light enough. It is wise to wear safety glasses while chipping to protect your eyes from flying slivers. For good measure, close your eyes as the hammer strikes. Using the edge on the flat end of the hammer break away small pieces from the edge of the stone. Gradually, one will get a feel for how much to break at one time. Chip the stone working toward the desired shape. Many a mason has formed a stone to a usuable shape only to take one additional hit - breaking it in the wrong place and ruining the work. Shaping stone is much like playing Black Jack.

Once one becomes experienced with shaping stone it will be easier to chip it to specification. A mason becomes sensitive to how hard a stone can be struck without smacking it to bits. One also becomes aware that breaking stone is an audial and tactile as well as a visual

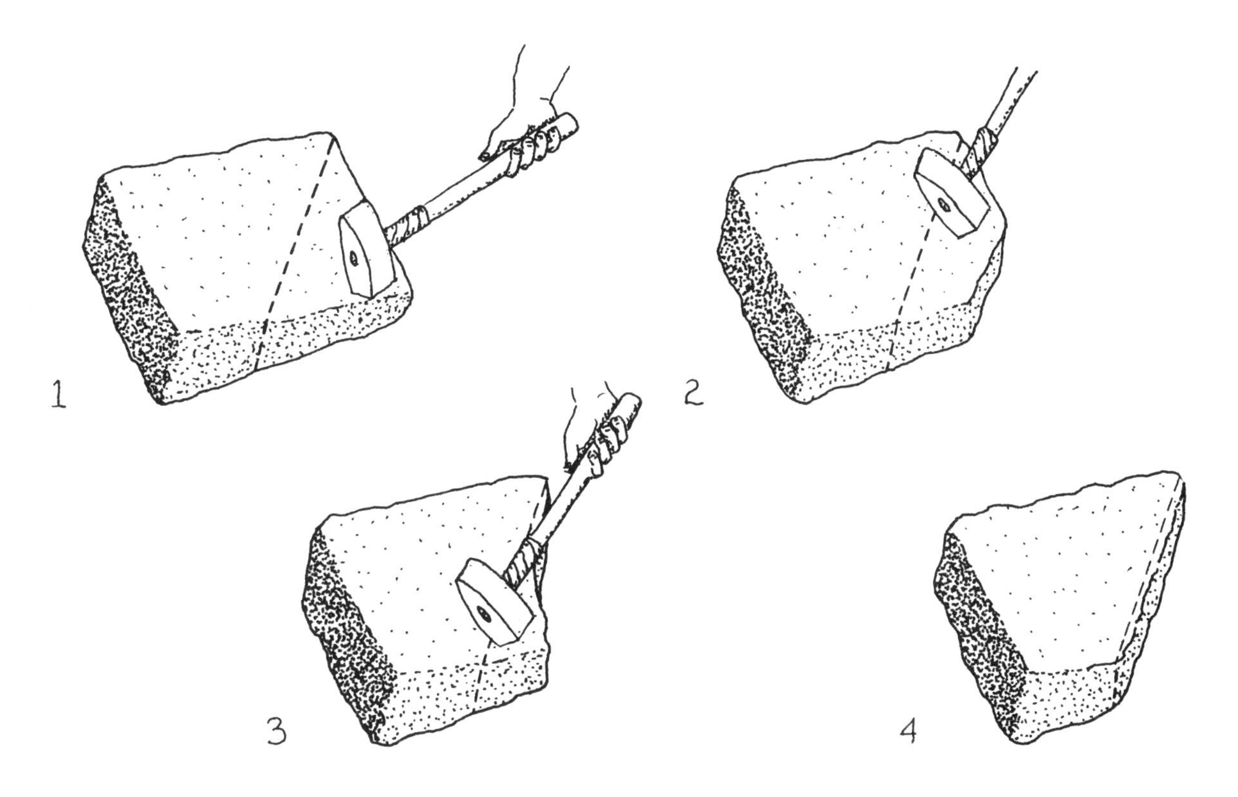

experience. When you chip stone, the sound of the hammer striking it changes from a sharp ring to a dull thud as the crack forms.

As it is held in the hand and struck with the hammer one may often determine by touch the way a stone will break.

The stone hammer can be used in other ways. Although it appears to be a simple tool each of the hammer's separate parts has a specific use. If stone is to be broken randomly into smaller pieces, use the full face of the hammer to achieve the greatest impact. The wedge end of the hammer is used to give a more specific and concentrated blow. The wedge is especially useful for splitting stone along its grain, much as one splits wood.

The stone hammer is the mason's most useful shaping tool but is not easy to find in a retail store. Most neighborhood hardware stores do not carry this tool, nor do many building supply houses. Write to Owner-Builder Publications for your free copy of our masonry tool catalog. Brick hammers, mash hammers and sledge hammers are fairly useful substitutes but it is best to use the tool designed for the job.

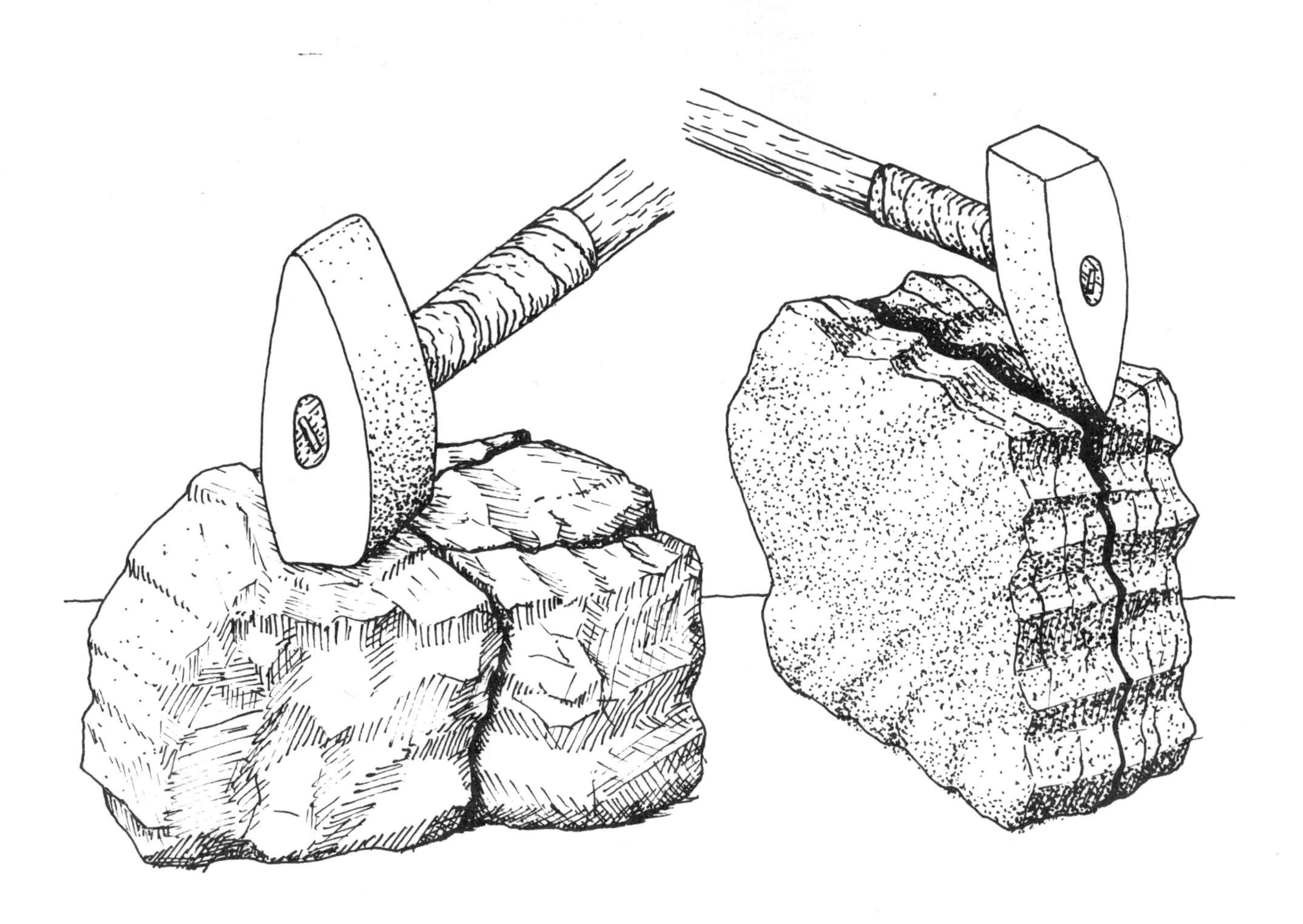

The stone hammer is the only tool needed to shape ninety percent of the stone a mason uses. It is the only shaping tool that a beginning mason must purchase. However, there are other tools that increase accuracy or come in handy for special circumstances. For instance, there are a variety of chisels useful when the hammer is not.

Suppose it is necessary to flatten the rounded top of a stone to provide a better surface on which to lay another stone. A hammer is mainly useful for breaking edges and not for working the surface of stone. A better choice of tool for this particular work would be a chisel. One type of stone-cutting chisel has a pointed end, called a point. This tool concentrates the force of its impact in a small area when smoothing surfaces. The fact that its impact is concentrated allows one to exert extra force in one place without concern for breakage elsewhere. When shaping hard stone, like granite, a point is about the only tool that will provide enough impact to be effective.

For smoothing softer stone the familiar cold chisel may provide the quickest results. It has a wider cutting surface but not the concentrated impact of a point. These chisels are useful for splitting stone along its grain. Both points and chisels come in a variety of sizes and can be ordered through a masonry catalog. It is wise to own two or three of each for they dull quickly.

Do not use the stone hammer to strike these chisels. Use a three-pound mash hammer for better control. When hitting the chisel with this hammer it is tempting to use as much force as possible. However, beyond a certain point hitting the chisel harder does not break stone better, it just dulls the chisel. It is best to proceed slowly and deliberately.

The metal of your tools is soft and flexible compared to stone which is hard and brittle. The brittleness of stone facilitates breaking it with metal but its hardness eventually dulls tools. The duller a tool the faster it continues to dull. A dull tool requires more force to do the same amount of work as a sharp tool. It is important to keep your stone-shaping tools sharp. However, a fine edge, like that honed for a wood chisel is too fragile for masonry and will collapse with the first impact.

There are other shaping tools for those who want to do very accurate work. Toothed chisels are a combination of a point and a wedge-shaped chisel. With this tool stone may be cut into almost any shape desired. Its fine teeth make it a fairly delicate tool so that its use should be restricted to soft material like sandstone, limestone or marble. These tools are difficult to sharpen and granite will dull their teeth immediately. When a toothed chisel is used one joins the ranks of sculptors and stone carvers.

There is one other useful hammer called a bush hammer known to some masons as a carlin hammer. It looks like a meat tenderizer and is used in relatively the same manner. This is the tool to use when smoothing curved, irregular or jagged surfaces. However, like

the toothed chisel it is a delicate tool and should only be used to work soft stone.

The extent to which stone is shaped is a matter of personal style. In this chapter, shaping stone has been presented largely as a utilitarian skill learned in order to make stone laying faster and easier. But some masons develop into skilled artisans who cut individual stones far beyond the utilitarian necessity. At this point stone-shaping becomes an art in itself.

Some builders prefer to keep stone in its natural form, the form in which they find it. But how does it evolve to this form? Creek stone is shaped by water; fieldstone is subtly changed by earth and air and quarried stone is blasted from rocky mountain sides. By the same token can't the hammer and chisel be thought of as vehicles of still another natural process?

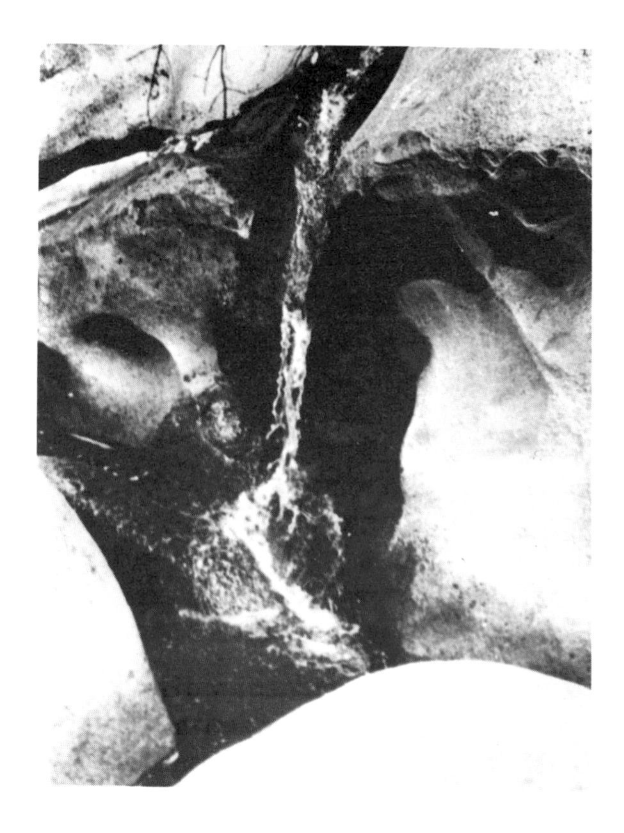

Mortar

Since stone construction was first used there has been one major problem, to fit stones sufficiently tight that wind, water or smoke may not pass through them. Without solving this problem the use of stone as a building material would be severely limited. What value is a shelter that does not prevent the entry of wind and rain or a chimney that will not contain smoke?

To solve this problem masons began using various plastic substances to fill the gaps between stones. Clay was a popular choice because it could be easily obtained. It was mixed with water to achieve a plastic state and smeared between stones. Clay will conform to gaps between stones and dries semi-hard. It is not permanent, however. It eventually washes away or flakes from between stones. Structures built with clay mortar need constant rechinking.

Eventually, mortar made of lime and sand was discovered and used. This mortar was a great improvement for it better withstood the action of water and wind. There are stone structures standing today with their original lime mortar still intact. This mortar is, however, not entirely resistant to water. It eventually washes away or creates dust because of its tendency to crumble.

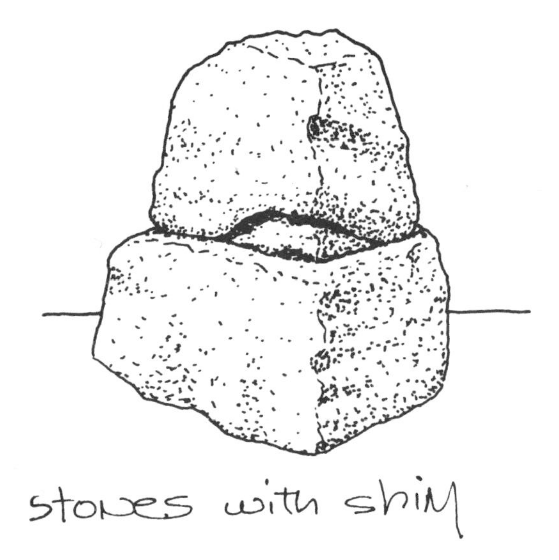

stones with shim

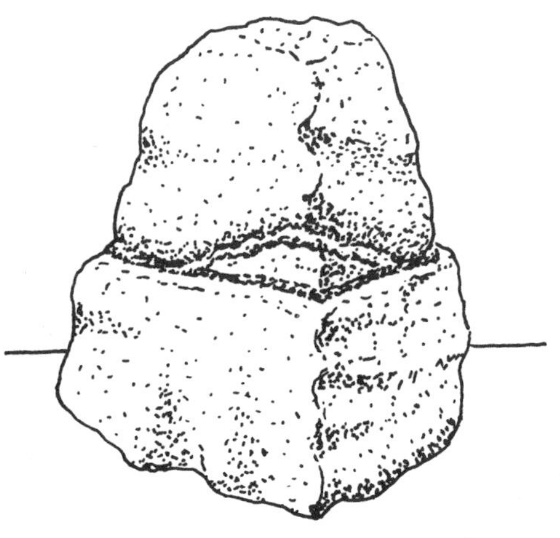

stones with shim & clay mortar

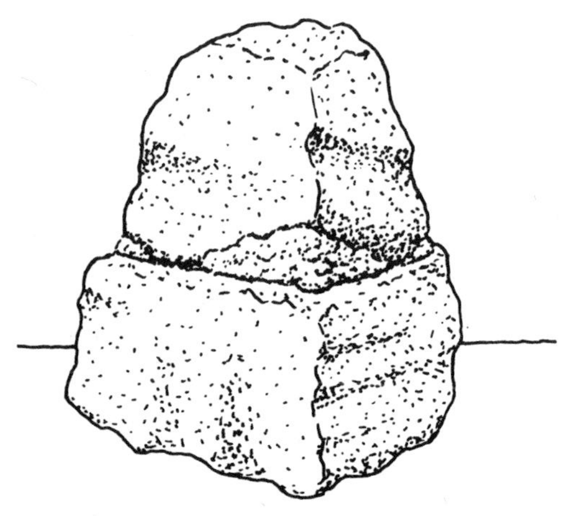

stones with portland cement mortar

It was not until the invention of portland cement, about a hundred years ago, that the mortar problem was solved. Portland cement is mixed with sand and water to form a permanent mortar. It sets hard, will not rapidly decompose and bears tremendous weight. On the other hand it is brittle, will not withstand bending, vibration or impact and is not waterproof.

The invention of portland cement changed the character of stone masonry. Filling gaps between stones was no longer a problem. Furthermore, the fact that mortar made with portland cement set up almost as hard as stone enabled it to be used as a structural material. Because of its compressive strength portland cement mortar will support the weight of stone.

To illustrate what portland cement meant to the practice of stone masonry we will observe a common building situation to see how the introduction of portland cement mortar affects it. Suppose you have two stones to be laid one atop another. When the upper one is placed on its bed it does not sit solidly; it wobbles. To solve this problem slip a shim where it will stiffen the arrangement.

If these two stones were part of a wall or chimney the remaining air spaces and cracks would have to be filled. If you were using clay or lime mortar it would be troweled around the shim before stone was laid on top. The top stone would then mash down, squeezing mortar into all spaces. A shim is still necessary because clay or lime mortar does not have the strength to hold the weight of stone.

If portland cement mortar were being used it would be applied in the same way, only in this instance the stone shim would no longer be necessary. Once portland cement mortar has set up, it will fully support the weight of stone. It will, in fact, be even stronger than a shim arrangement because the upper stone will be supported at all points. Stones in dry walls usually touch only at three or four points. In other masonry constructions it is even possible to lay such a thick bed of mortar that stones do not touch at any point.

Portland cement mortar fills the gaps between and permits a solid, stable fit of one stone on another. It is important, however, to learn what this mortar will *not* do. It will not bind stones together. Mortar is not glue and the bond between two stones is not strong. Even after the mortar has fully set it is often easy to pick the top stone off the bottom one. This means that portland cement mortar cannot be relied on to defy gravity. A pier built like the one shown will be unstable - even though the stones are mortared.

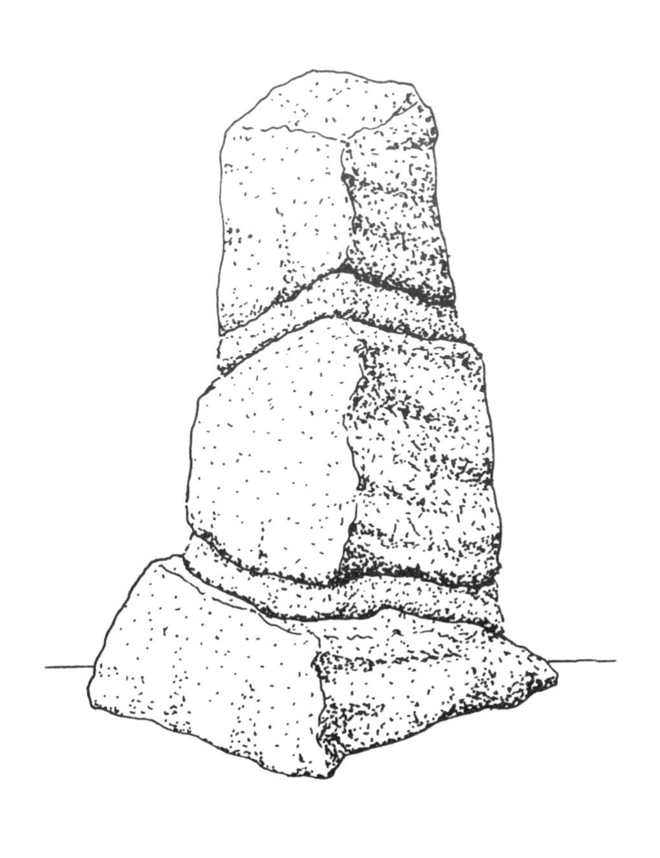

stable

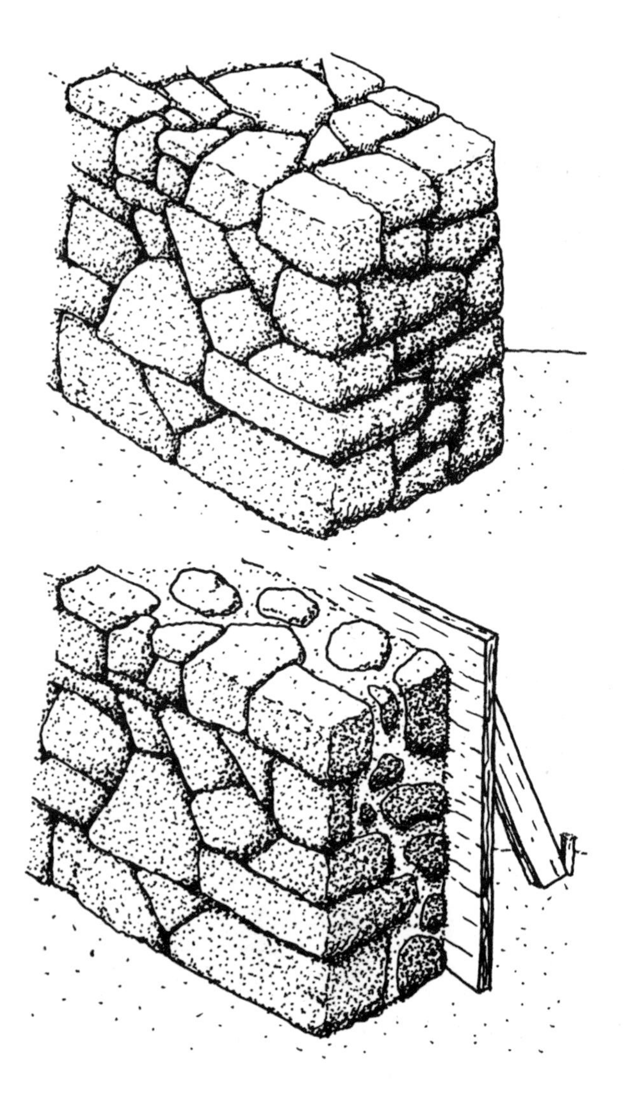

The strength of portland cement mortar can save a person much time and effort when building. Here are cross sections of two foundation walls. One was built dry and the other with portland cement mortar. From the front, these walls look identical although they were built differently. The wall above was built by the dry wall method. Each stone was carefully selected, placed and shimmed. This wall is strong, stable and durable.

The bottom wall was built against temporary backing. Identical stone was laid without mortar in the face of this wall to achieve a dry wall effect. The rest of ths wall, however, was built with mortared rubble packed behind the face stone. This wall is as strong as the dry wall, thinner, used less material and was quicker to build. It does not, however, have the same integrity or craftsmanship of a dry wall.

Built with the aid of mortar, all the stone in the above example is laid in accordance with the two rules of stone masonry. As long as the principal structural material is stone these rules *must* be followed. Although portland cement mortar is strong it cannot accomplish what gravity does not allow.

Look at the cross section of this foundation wall. Here mortar is used purposely to circumvent gravity. None of the stone in this wall is laid in compliance with the structural rules of masonry. In fact, none of the stone is this wall is structural. It is mortar alone that holds this wall together, not stone. Without mortar the stonework would collapse.

Walls like this are found everywhere. They are not stone but rather concrete walls with pieces of stone stuck in for decoration. If built well a concrete wall will last a long time. You will, however, often find these walls in various stages of deterioration. Inasmuch as face stones are merely embedded in mortar and do not support one another they often loosen and fall out, leaving hollow spaces. It is tempting and easy to misuse portland cement mortar when building with stone. It must be used with discretion.

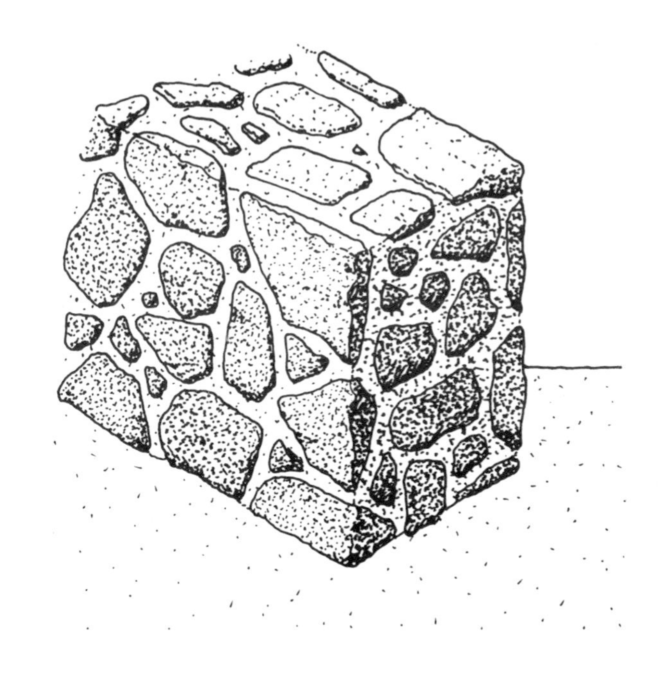

To build properly with this mortar one must understand how it works. Portlant cement is a powdery mixture of gypsum, lime and clay that has been fired. When water is added to this mix a chemical reaction starts in which the various elements combine into a homogeneous mass. In this process cement coats the particles of sand with which it is mixed, binding them to create mortar. If you wish to make concrete, add gravel which will be coated by the sand-cement mixture. The gravel mixture may in turn coat larger stone added for fill. The portland cement is the binding agent while the aggregate adds strength.

Portland cement does not dry; it sets. That is, when the chemical process is completed and the cement-water mix has properly combined it becomes solid. This process takes place more rapidly when weather is warm but more efficiently when it is cooler. Water is often sprayed over the work for concrete must not be allowed to dry before the reaction is complete. This is why curing cement must be kept damp for several days.

Freezing temperatures halt this chemical reaction by turning the water in mortar to ice. Cement should not be used when there is danger of freezing. Also too much water dilutes mortar and again prevents proper setting. The wetter your mix, the weaker the concrete. Rain will wash wet cement out of sand. Once

you understand these basic facts about portland cement you can better control its use.

There are a number of different proportions for mortar. Basically, they all use the ratio of one-part portland cement to three-parts clean sand. Some masons prefer to add a little lime or fireclay, giving the mortar a stickier, more plastic constistency. You will determine the mix appropriate to your style.

Mortar can be mixed in a wheelbarrow, mortar box or cement mixer. If you mix manually, to avoid back strain it is wise to purchase a mason's hoe, the type with holes in its blade. A mixer saves even more time and effort. First add dry ingredients and mix them thoroughly. Then add water as needed. Mortar should be firm yet plastic. It is very easy to add too much water. At one moment the mix may seem quite dry, but with the addition of a little water it may suddenly become thin and runny. If this should happen you must add more sand and cement.

The ease of handling mortar depends on its consistency. If it is too dry it will be hard to spread and will not stick to stone surfaces. If mortar is too wet it will not support the weight of stone. It will also tend to run over the face of the wall, creating a messy appearance. Once you have worked with a satisfactory batch of mortar, called "mud" by masons, you will understand its advantages and become particular about mixing.

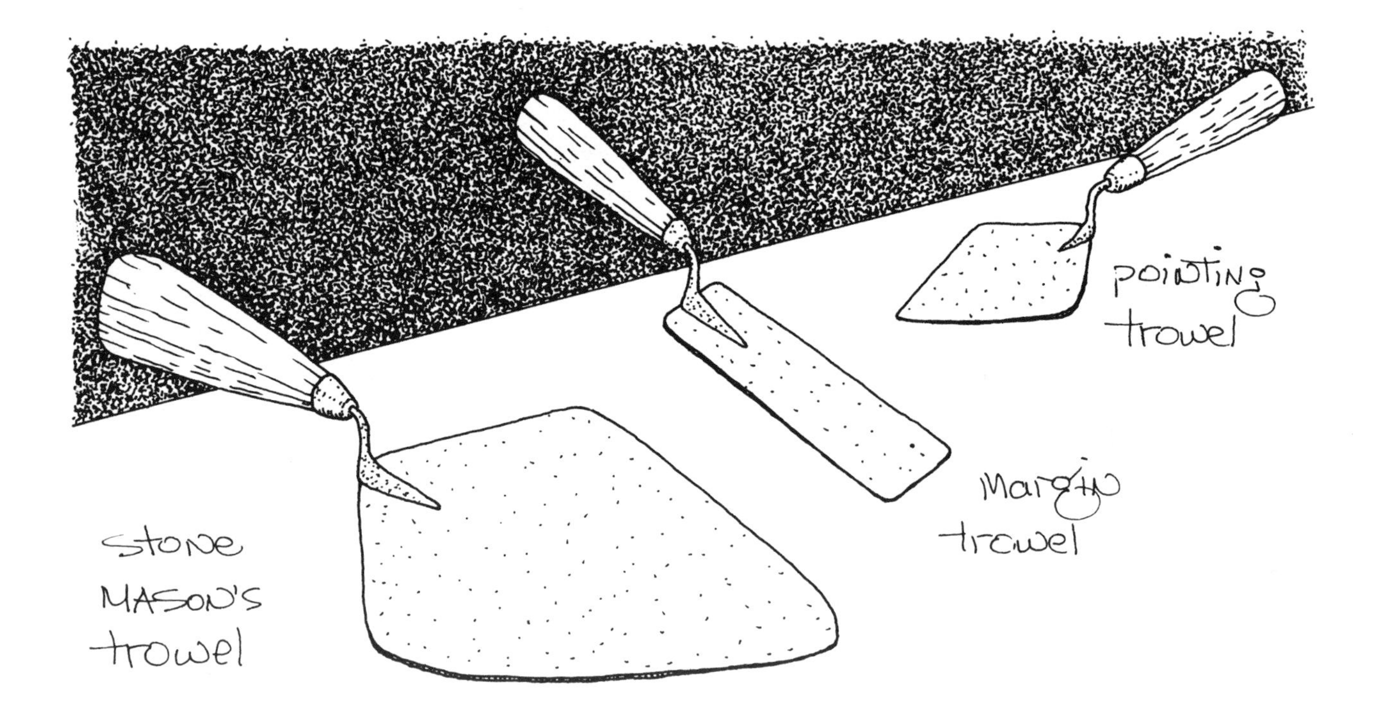

The procedure for building a mortared wall follows: Select the stone you are going to use in a particular space. Do any shaping or cleaning that is necessary to satisfy yourself that the stone is entirely ready to lay in place. Only after the stone is ready do you trowel down (spread) a bed of mud (mortar) in the space.

Mixed mortar is most conveniently stored for use in a wheelbarrow or mortar tub. Plywood mortar boards tend to dry mortar too quickly, and it is hard to dip a trowel into a bucket. To handle mortar you will use a trowel that tooks like one of those illustrated here. Stonemason's trowels tend to be shorter and rounder than bricklayers'. Smaller trowels may also come in handy for packing mortar into crevices impossible to reach with bigger trowels.

Apply mortar with your hands as little as possible. Mortar ingredients irritate and dry out one's skin, creating painful cracks. The abrasiveness of sand can rub skin from the tips of one's fingers. Some people say they get a better feel for their work by applying and removing mortar with their hands. If that is true for you wear a pair of rubber gloves.

When you trowel mortar use only as much as necessary to provide the bed with sufficient covering. Too much mortar will only squish out and over the stone face. Do not trowel smooth the mortar; let the stone mash it down. In this way gaps will more certainly be filled. Once a stone is laid in place try not to move it. Any movement will weaken the bond between stone and mortar. Also taking up stone creates a cement-staining mess. After a stone has been placed in the bed of mortar clean off the excess oozing from joints. Some mortar may have to be pushed into unfilled spaces between stones while also filling vertical joints.

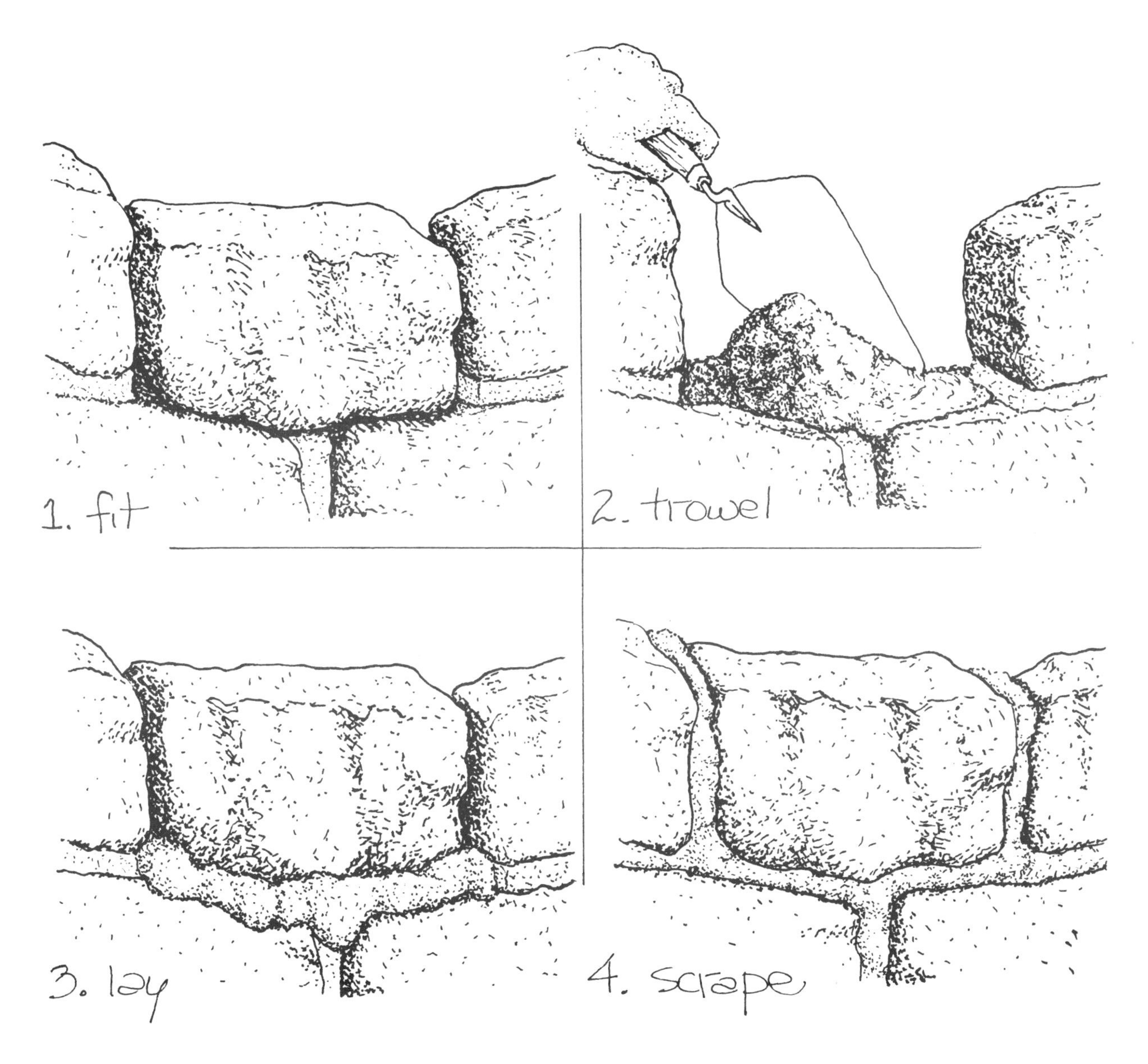

From the moment you lay a stone in its bed mortar begins to harden at an accelerating rate. The stone absorbs water from the mud, causing it to dry out. Eventually, the consistency will change from wet and plastic to dry and granular but not yet hard. On hot days this may occur in an hour or less. On cooler, wetter days it may take overnight. When mortar reaches this granular state it is ready to be finished or pointed.

The neatness with which mortar joints are finished can make the difference between an amateur or a professional-looking wall. If the joints are smeared and sloppy then the whole wall looks that way, no matter how well the stone is laid. On the other hand, a wall with neatly dressed joints will look trim even when there are large gaps between stones.

Mortar joints can be finished in a variety of ways. One method employs no visible mortar. This is achieved when stone is laid dry and is backfilled with mortar. If stone is tight mortar will not ooze to the front of the wall. When it does, excess mud can be trimmed away.

Many people, however, prefer the more substantial appearance of walls with visible mortar joints. The dry wall look has been criticized for its deep crevices that contain freezable moisture, dirt and insects. Of the many ways to finish exposed mortar, the simplest involves waiting until it has taken on a granular texture. It is then scratched and brushed until the joint is recessed. This creates a desirable shadow effect around each stone, enhancing its character.

The two tools you will use for this operation are the long, thin pointing trowel and the stiff-bristled brush. There is no brush made specifically for this purpose, but the brushes sold in auto supply stores for cleaning engine parts work well. Wait until the mortar has set up sufficiently, and then use the pointing trowel to scratch excess mortar from between the stones to a desirable depth. How deep depends on your taste. Some masons prefer mortar joints that are deeply recessed while

others bring mortar to the face of the wall. The brush can be used after the trowel to smooth mortar and give it an even texture.

Another way to finish mortar joints is seen more in formal architecture than in owner-built homes. Here, joints are finished so that they extend beyond the surface of the stones, outlining each to give them a precise, framed look. This method requires much skill and practice. Work attempted in this style looks sloppy when done with the wrong tools. Masonry contractors who offer this style generally have a separate pointing crew who commence work when stone laying is finished. These joints are applied with a special tool and use a rich mix of mortar containing additional cement.

There are numerous variations of these methods for finishing mortar joints. Some builders prefer to accentuate mortar while others conceal it. There is a technique in which the mason smooths mortar across the surface of a wall, virtually hiding the face of some stone. Still another approach allows mortar to squeeze out between stones, recording its former plastic state as it hardens. You may discover yet another style of your own.

Once the wall is finished, the joints pointed and the mortar set and well cured, it is likely that faces of the stone will be smeared with

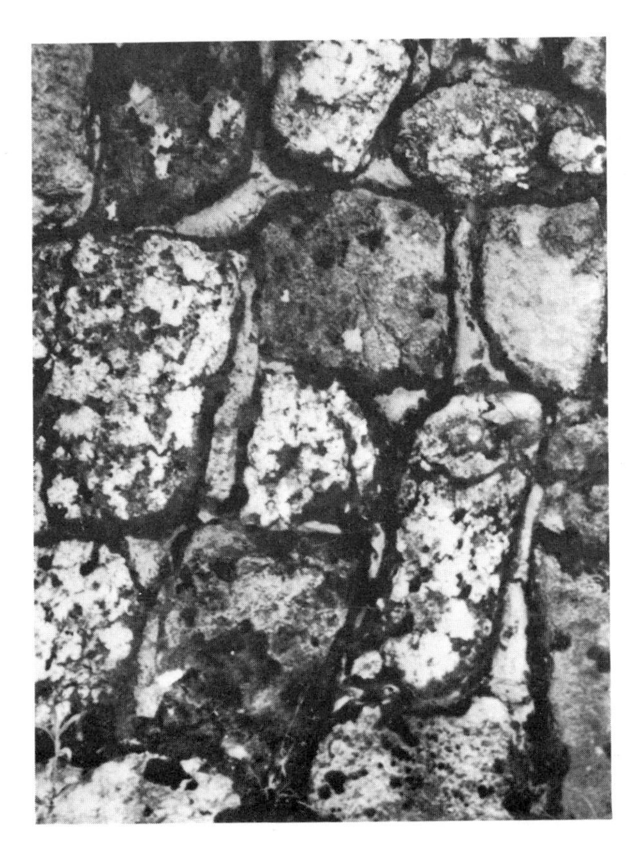

excess mortar. Pieces of mortar can be cleaned away with a hammer or trowel but the stains and smears must be chemically removed. Mortar stains are cleaned from stone with a solution of muriatic (hydrochloric) acid and water. The acid can be bought at a building supply store. When you get it the acid will be full strength. It should be diluted with water: one part acid and up to seven parts water. The proportion depends on the job you have to do. When acid is applied to mortar smears it dissolves the lime, disintegrating the cement. This acid will not, however, decompose the surface of most stone. Apply the acid with a long-handled scrub brush. Keep your skin and eyes well protected because this acid is extremely caustic. It will also destroy clothing.

Wash the wall thoroughly to clean off any dirt and prevent streaking. Then apply the acid solution. When the acid bites into cement it fizzes and bubbles in a dramatic display of its potency. Allow it to remain on the surface several minutes and then rinse it with clean water. If cement stains are still visible try a stronger mix. The acid, of course, dissolves the surface of mortar joints as well as the mortar smeared on the face of stone, but it does not harm the structure of the wall. Stronger mixes tend to antique mortar joints, making them pitted.

Most masonry done by an owner-builder requires mortar. The introduction of portland cement mortar has changed stonework considerably over the past century. Stone is now used in ways that were previously impractical. Both stone-faced and stone-formed walls were unknown before the advent of portland cement. When using portland cement, however, the owner-builder must be careful not to overestimate its ability to bind stone. Like other types of mortar portland cement also deteriorates faster than stone, therefore the primary structural element of a masonry wall should be stone itself. The role of mortar is mainly to fill gaps.

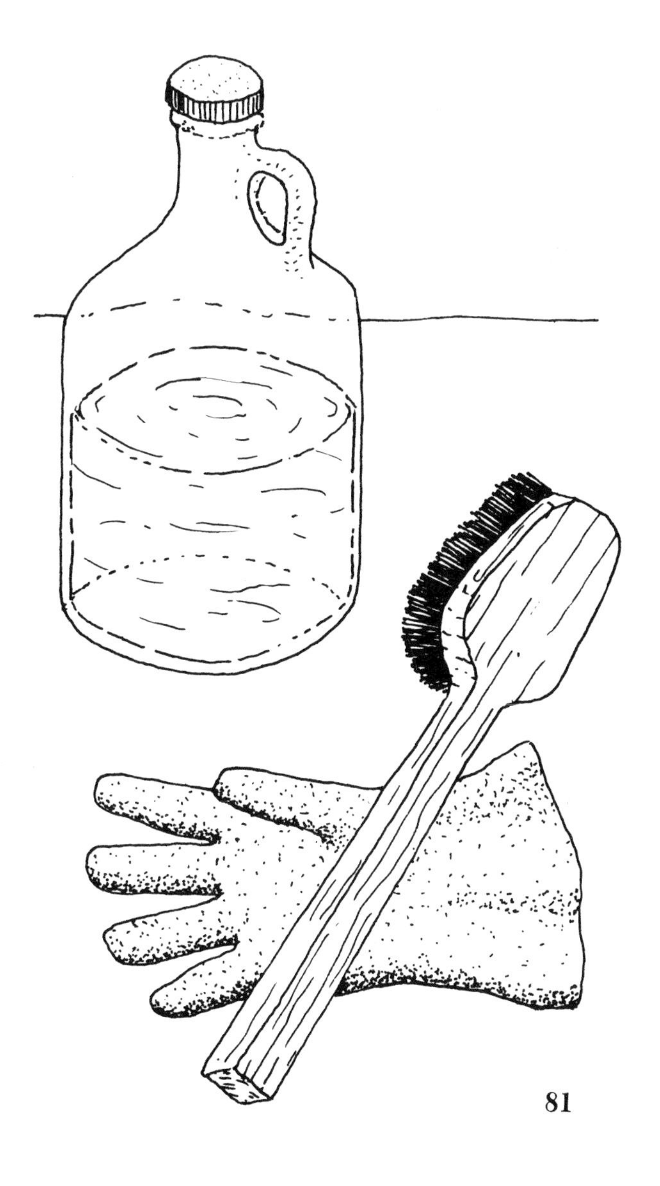

Methods

Laid Masonry

Until recently all stone masonry was laid masonry. The pyramids, the cathedrals of Europe and the stone barns of America were all built this way - the most obvious, direct and simplest method of construction. Using this approach, the builder needs only the aid of a few basic leveling tools when placing stone upon stone to create a solid wall. Yet the very simplicity of this method also means that it requires much patience, care and skill. Unlike other methods of laying stone, the mason does not have the aid of an existing wall against which to build nor does a form hold the stone in place until the mortar sets up. The owner-builder who selects this approach must rely completely on the way he places stone for the strength and form of the wall.

The skill and care required by this method should not discourage the owner-builder from using it. The very simplicity of laid stone masonry makes it a natural process for a person of limited experience. The primary resources required are time and patience.

Laid stone construction can be used for many types of structures: fireplaces, chimneys and foundation walls, to mention a few. This chapter will focus on building walls for a stone house to illustrate the basic elements of this method. Constructing house walls poses a number of considerations. Before starting to build, much planning needs to be done. Questions about the wall's size, thickness, insulation and the placement of windows and doors must be answered.

Dimensions

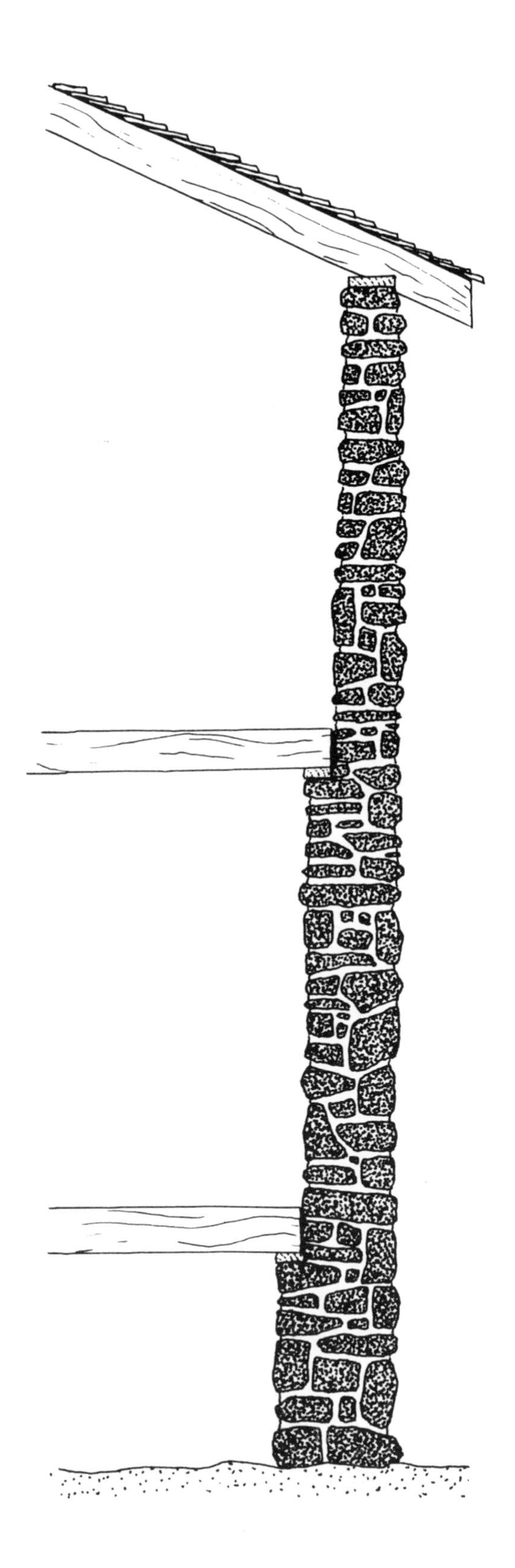

How high can a wall be? How thick need it be? These are questions whose answers are subject to a number of variables such as the design of the house, the quality of the stone and the skill of the mason. Some houses demand more from a stone wall than others. For example, a house with a shed roof needs only vertical support. A house with a gable roof, however, puts outward pressure on walls and requires more lateral strength.

The quality of the stone used is also a consideration when planning a wall. One built with flat, square stone will be stronger than a wall built with rounder stone. Finally, the skill of the mason is a variable. The way stone is laid determines the integrity of the wall.

Questions of convenience are equally important when you consider building a stone wall. You will find that most owner-built stone houses are one story. This is not because two-story walls would not be strong enough, but the work and difficulty of laying stone increases proportionally with their height. Scaffolding has to be erected and stone and mortar carried higher. In addition, the wall must be constructed thicker at the base to accomodate the greater height.

How thick should a wall be? In general, a one story wall must be about a foot thick. If it is two stories high its thickness needs to be increased to sixteen inches while the second story wall may be only twelve inches. This suggestion is once again subject to many variables. For instance, it may be that the size of the available stone in your area will make it difficult to build a double-faced wall just twelve inches thick. It may prove easier to build a one story wall a full sixteen inches thick even though this is not structurally necessary.

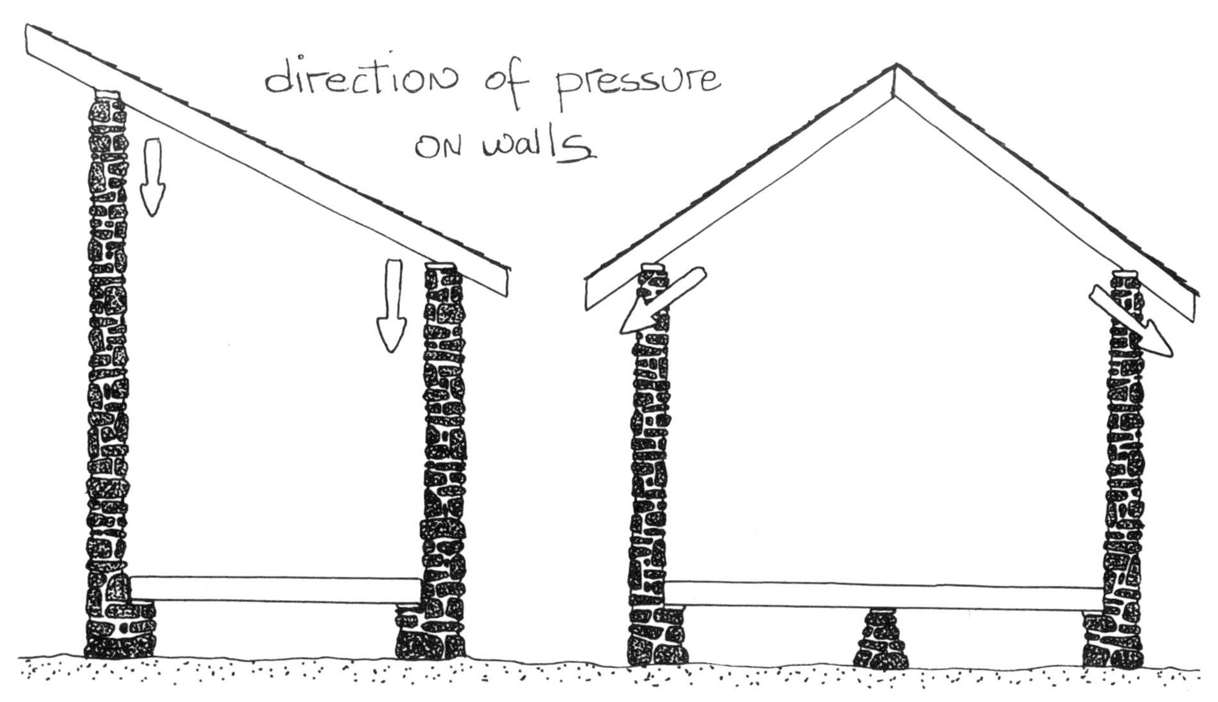
direction of pressure
on walls

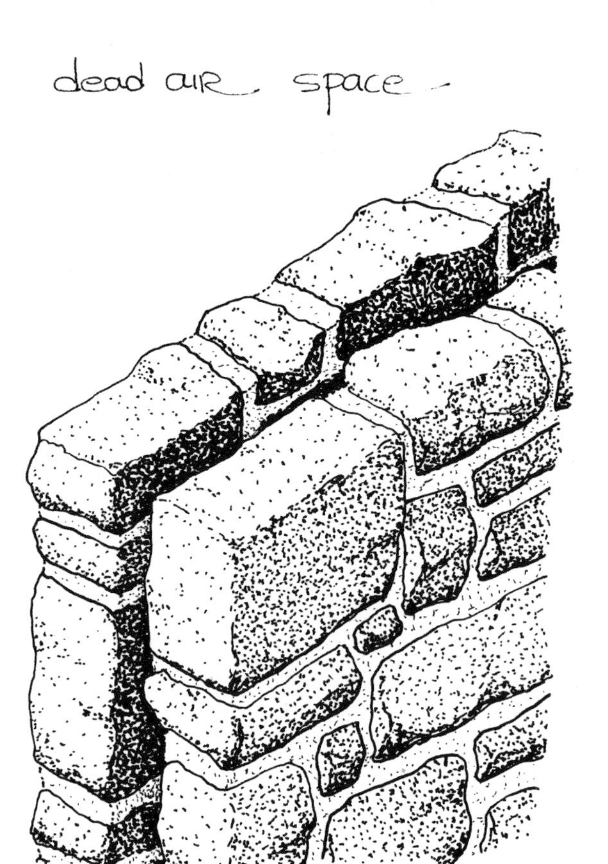

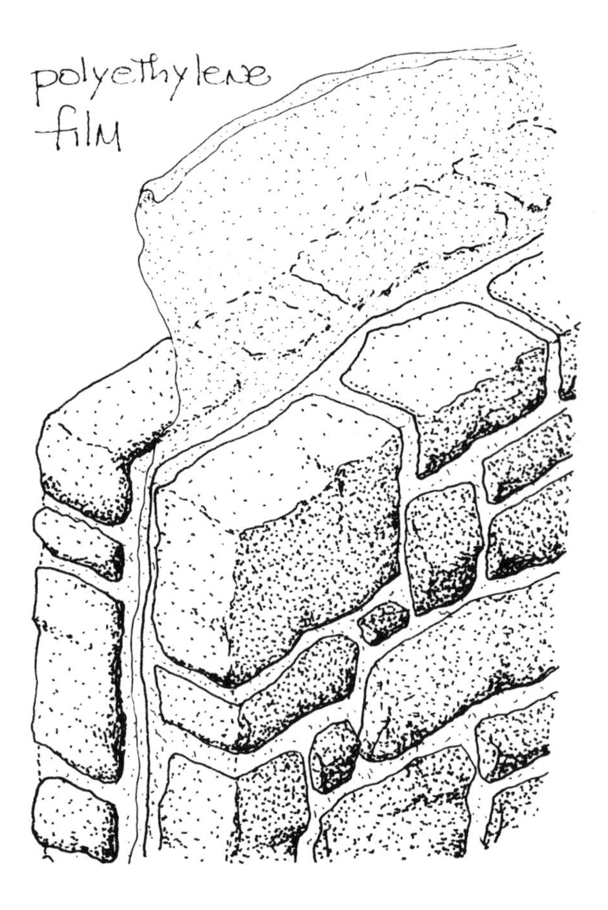

Insulation

The insulation of walls is an important consideration when designing a house. The reader may be surprised to learn that a solid sixteen-inch stone wall provides only one-tenth the insulative value of a wood-framed wall containing three-and-a-half inches of fiberglass batt insulation. Increasing the thickness of a stone wall, without adding other materials, will not improve its insulative value appreciably. One reason for this is the high thermal conductivity of the uninterrupted mass of stone and mortar. Another reason involves the relatively porous quality of stone and mortar which allows the penetration of

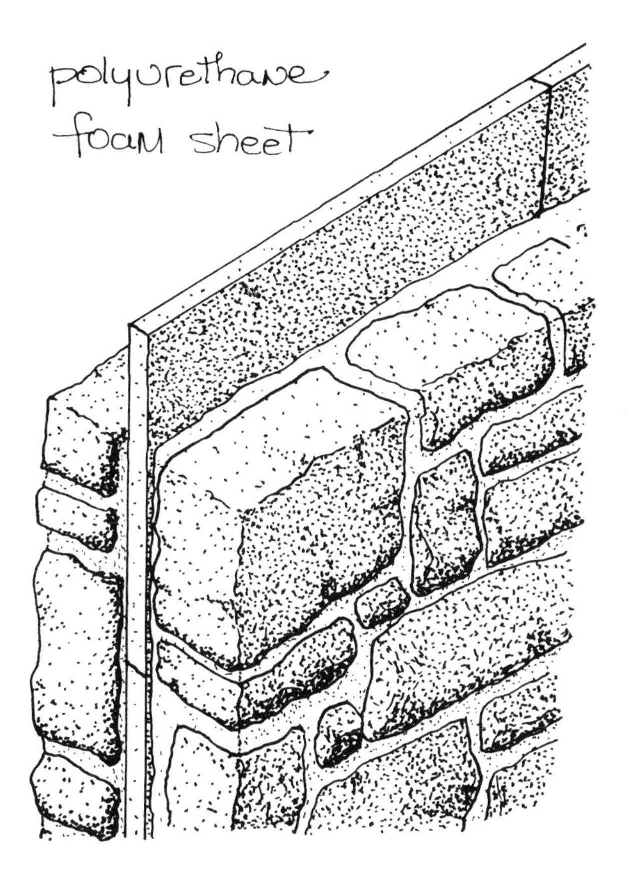

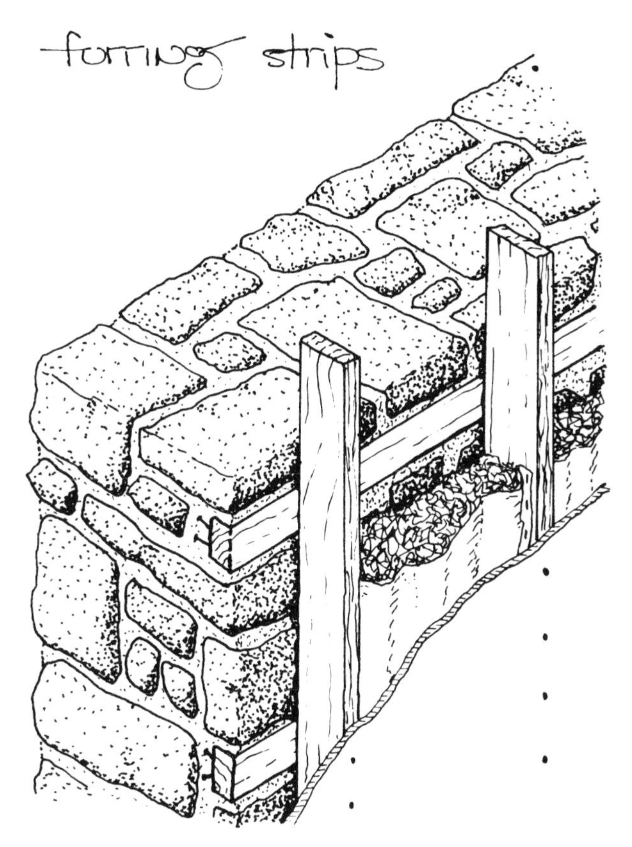

moisture, bringing with it the heat of summer and the cold of winter. The opposite effect is also true; interior heat will escape outward through walls with comparative ease. This means that, if one is planning a house with exterior stone walls, extra provisions must be made for insulation.

There are a number of ways to increase the insulating quality of a stone wall. The simplest way is to include a dead air space between the two faces of the wall. This air space will impede the movement of heat and cold through it. Another method is to use polyethylene film (plastic sheeting) in this wall space. It excludes all mositure but still allows thermal conductivity. A more effective method, offering protection against moisture and thermal effects, includes the use of a sheet of polyurethane foam between the two faces. A two-inch sheet of this material will insulate almost as well as three-and-a-half inches of fiberglass batting.

All these insulating methods have the disadvantage of dividing the wall, thereby, decreasing its strength. There are a number of ways to tie the two halves of a wall together. These will be explained later. One additional method of insulating a solid wall does not compromise its integrity. With this approach wooden furring strips are laid in the mortar in the interior side. After the wall has been completed, these strips are used as nailers for paneling interior surfaces. Fiberglass insulation may be installed between the stone and paneling. This method is effective but has its disadvantages. It covers the interior stone surfaces with other material and creates additional work. It would be just as easy, or easier, to build a stud wall first and face it with stone later. Extra care must be taken that the interior wall is built flat and straight, that furring strips are placed on a plane with one another and that they are well anchored by nail heads protruding into the mortar.

Wiring & Plumbing

The builder may also want to include electricity and plumbing in stone walls. This is not easy and can often be avoided by using interior wood-framed walls to accomodate these needs. Outlets can also be installed in floors rather than in walls. If it is necessary to have outlets, switches or plumbing fixtures installed in a stone wall, it looks better to have them built into the wall rather than attached to it afterward.

All wiring in a stone wall must be protected by electrical conduit as mortar deteriorates other types of insulative covering. The conduit and outlet boxes can be built into the wall and wires may be laced in afterwards. This should be planned ahead of time.

Water pipes, both copper and plastic, can be mortared into a wall. When assembling these pipes it is prudent to test their couplings beforehand since, once they are embedded in wall, a leak will be virtually impossible to repair.

Other Factors

Several other details of a stone wall must be planned ahead of time. The size and shape of doors and windows should be decided and their place in the wall determined prior to construction. Although changes can be made later it helps to have some idea of their placement.

Be sure to plan how floor joists and rafters will be attached to the wall. If a stone wall meets a framed wall, how will they be joined? Such problems can often be solved by setting wooden nailing blocks into the mortar at predetermined places. Sometimes studs are set in mortar where the stone wall joins the wooden wall. Cabinets and shelves can also be attached to masonry by nailing them to similarly set-in blocks.

Wood in contact with masonry should be treated with preservative. Masonry attracts moisture and tends to rot the wood it surrounds. Wooden blocks may be attached to the wall with nails which embed themselves in the mortar. Without this insurance the wood may shrink and pull out of the masonry.

A more effective way of attaching wood to masonry is to sink bolts in wet mortar. Wood is then drilled and fastened in place after mortar is set. This is only necessary, however, where the joint is likely to undergo stress. An example of where bolts may be used is when a wooden plate is attached to the top of a masonry wall to accommodate the nailing of rafters.

Footings

Once the wall is well planned the next step is to prepare the site for building. In this section there is no mention of laying out the house because it is adequately covered by most house building manuals.

Before the stone wall can be built an adequate footing must be prepared to support it. To understand the necessity for a footing one has only to look at a cracked wall or tilting chimney that were built directly on the ground. If a footing is not used and the base stones lie immediately on the surface, several things may happen. If the soil is soft or unstable, it may in time sink under the weight of the wall. Often roots and other organic matter beneath the wall will rot, allowing the ground to give way in a particular spot even more damaging to the wall. Running water can wash ground right out from under a wall. And even if the ground is solid, it still may heave with freezing.

To avoid these potential disasters dig a trench. It must go below most organic matter and top soil into solid ground. It must also extend below freezing depth. In southern and western regions of the United States this depth is only a foot. However, the northern states need trenches four or five feet deep to get below the frost line. To determine the depth of the frost line in your area, consult a local builder.

The term "footing" is very appropriate. A well made footing supports a wall, distributing its weight over an area greater than the wall's base — just as our feet support our legs. Generally the footing should be half-again as wide as the wall. Exact dimensions depend on the stability of the soil.

When digging the trench for a footing, keep the sides smooth and regular. Cut back all protruding roots. The base should be flat and smooth. A footing with a wedge-shaped or rounded bottom will sink more readily than a flat one. Be sure to remove all loose dirt from the trench. Do not attempt to pack it down for you will not be able to compress it adequately.

Always dig down to solid, undisturbed ground. If you are working in an area that has been recently filled, dig below the fill dirt to solid earth, even if the resulting trench has to be very deep. It takes years for filled areas to settle enough to support the weight of a wall.

Footings are usually made of poured concrete but this is not always necessary. The advantage of concrete is that it provides solid continuous support for the wall. If one digs to bed rock or hard ground, a solid concrete footing is not required. One can immediately begin building with stone and mortar on these surfaces.

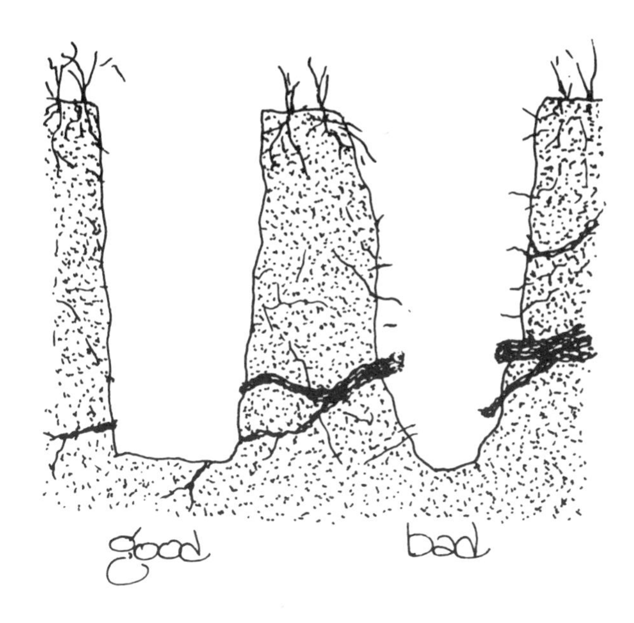

However, if you do not hit solid ground or rock, the support provided by a concrete footing is mandatory for stabilizing the wall. The thickness of poured concrete footing should be from eight to twelve inches for a stone wall and even thicker for a chimney. Sometimes where the ground is exceptionally soft or saturated with water, it may be necessary to reinforce the concrete with reinforcing rods. The steel gives concrete a tensile strength that it does not ordinarily have.

Once you have dug the trench, measure it so that you can determine the amount of concrete needed. If it is over two cubic yards, consider having ready-mix concrete delivered in a truck. Mixing is slow, exhausting work. It takes a person the better part of a day to mix and pour two cubic yards. With a truck of ready-mix the job can be completed in half an hour. Depending on the distance the truck has to travel to reach the site, it may not cost appreciably more for ready-mix than for the raw materials you would have to mix yourself.

If ready-mix is used, make sure that the access road to the site is in passable condition. Make certain the truck can be driven close to the footing trench. It is easier if the truck can back to the site so that concrete may be poured directly into the trench. If this is not possible, it will be necessary to carry the concrete from the truck to the trench in wheelbarrows. If this is to be done, it will be an

excellent time to enlist the help of any neighbors who owe you work. Several manned wheelbarrows make the job much easier. A wheelbarrow full of concrete is heavy and unwieldy. Prepare a clear path and avoid the situation where you will have to push a full wheelbarrow uphill. Even attempting to hold a wheelbarrow back while going downhill is difficult.

If ready-mix service proves impractical, you will have to mix concrete by hand. For a job this size you will need a mixer. There is an active market for used cement mixers. You may find one and resell it when the job is over. An electric-powered one is preferable; it runs more quietly and dependably.

For most footings, concrete can be mixed at a ratio of one-part portland cement to three-parts sand and four-parts gravel. Add enough water to moisten the mixture. A sandy-textured mix is stronger than a soupy one. As you pour each batch into the trench, much time and money can be saved by throwing in unsuitable building stone for fill.

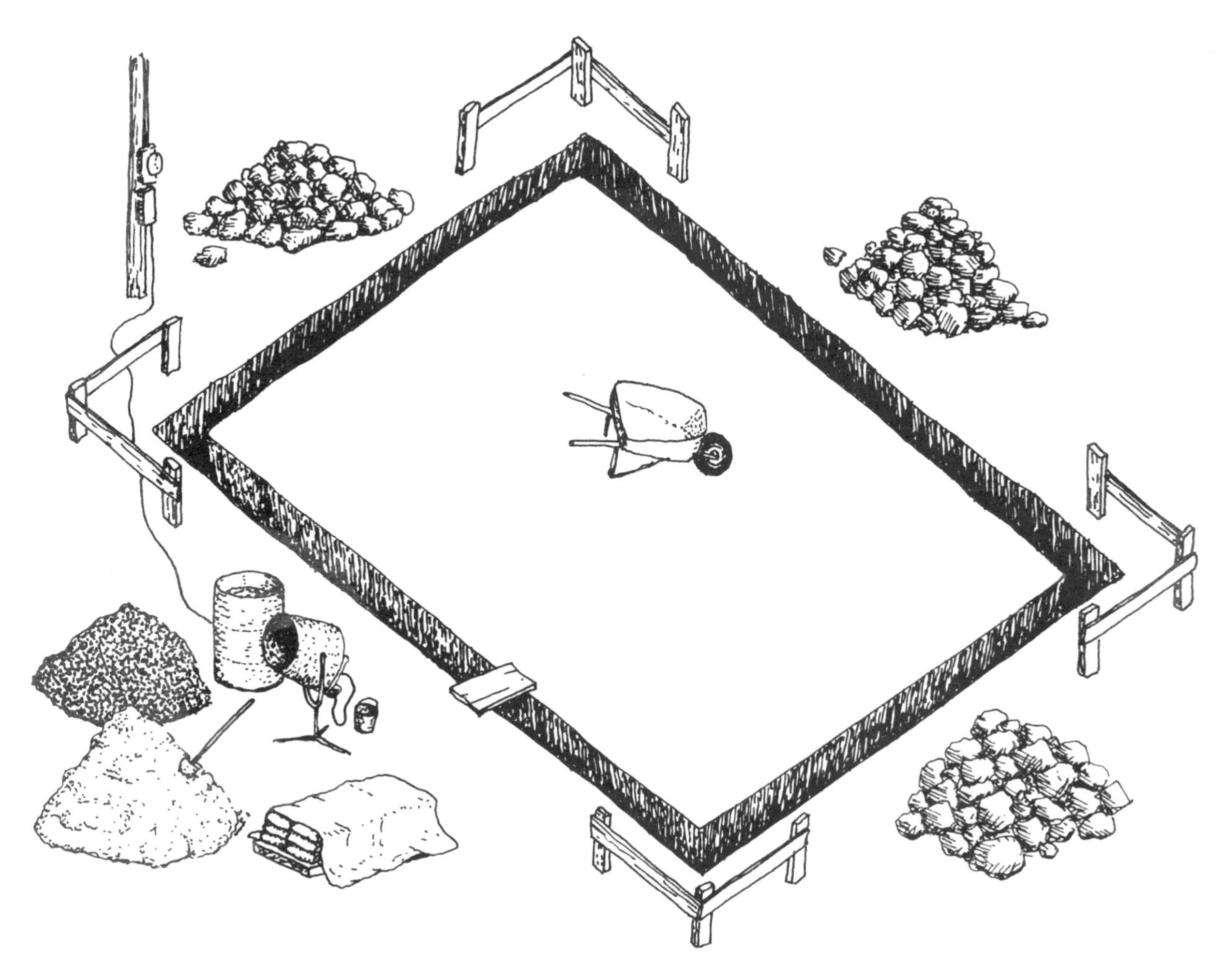

Once the footing is poured to the desired depth, smooth it with a trowel or a block of wood. Since you are working with stone, it is not crucial that the surface of the footing be either level or smooth. Some masons prefer to lay their base stones directly into the newly poured footing. This insures that the base of the wall is well anchored.

In colder climates after an adequate footing is poured, there may still be a distance of several feet between footing and ground level. This space could be filled with concrete but this may prove expensive. Although it is not easy to lay stone several feet below ground this is another alternative. Of course you do not have to concern yourself with how the finished product will look. It is, however, important that all stones sit firmly on one another, even if there are large gaps that must be filled with mortar or concrete.

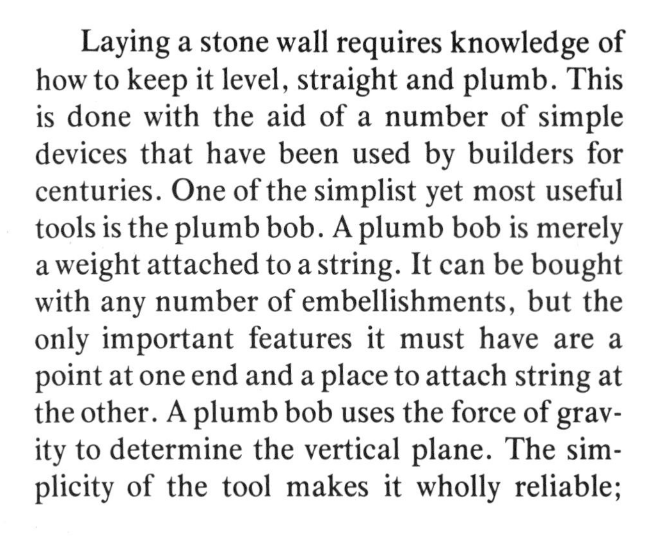

Leveling

Laying a stone wall requires knowledge of how to keep it level, straight and plumb. This is done with the aid of a number of simple devices that have been used by builders for centuries. One of the simplist yet most useful tools is the plumb bob. A plumb bob is merely a weight attached to a string. It can be bought with any number of embellishments, but the only important features it must have are a point at one end and a place to attach string at the other. A plumb bob uses the force of gravity to determine the vertical plane. The simplicity of the tool makes it wholly reliable;

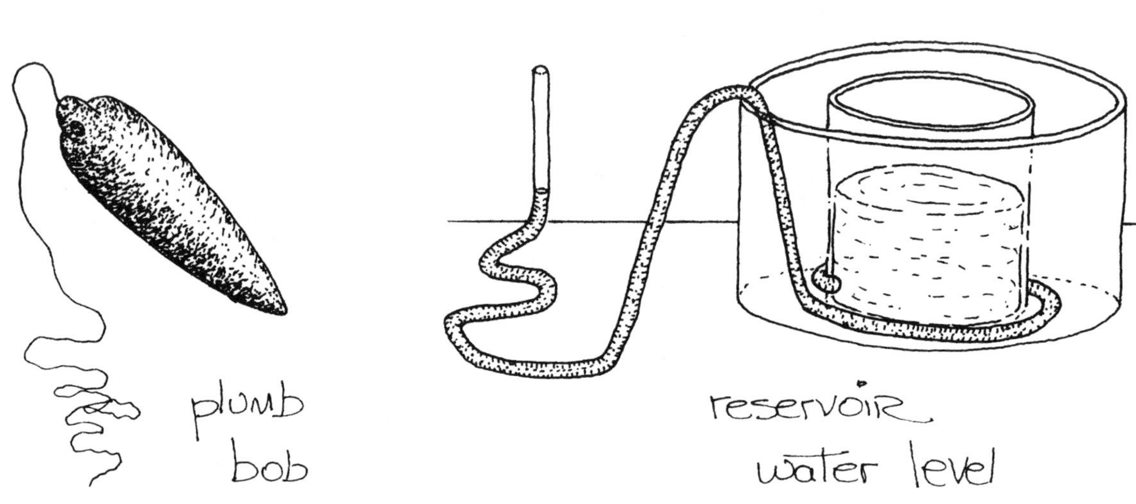

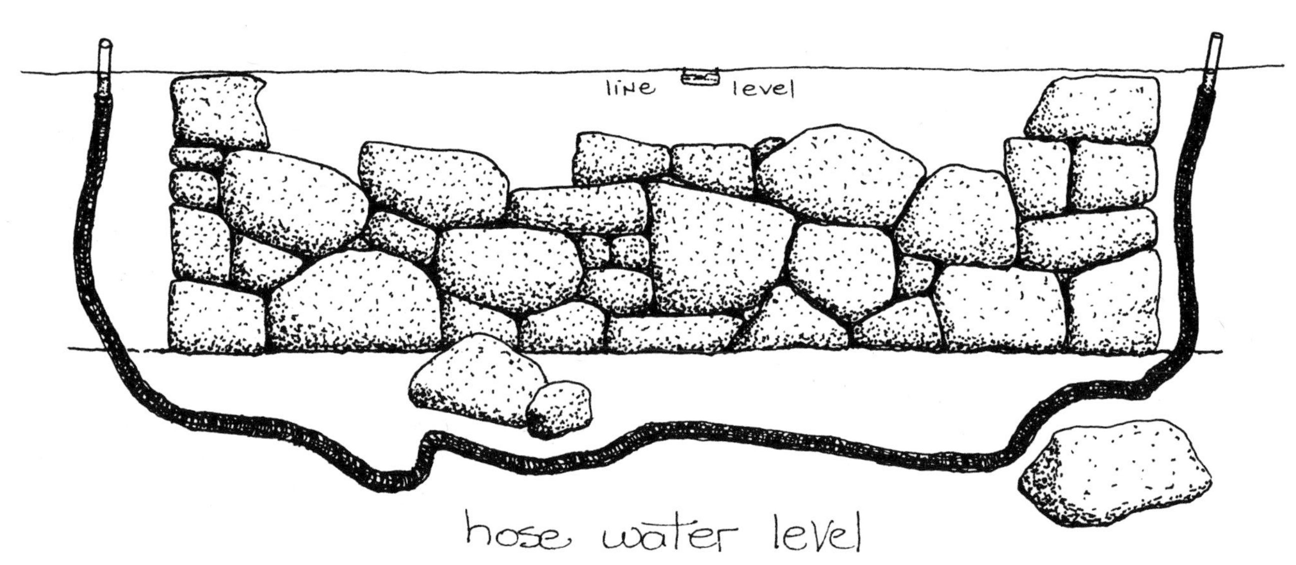

nothing can go wrong with it. The plumb bob can either be hung beside a wall or held at arms length to eyeball its straightness. It is also useful for setting more permanent guides for wall building. This will be explained in more detail later.

An equally simple tool is the water level, a hose filled with water. It takes advantage of the fact that the reading at one end will always be level with that at the other. Store-bought water levels are available with a central reservoir; they are easier to use and can be operated by one person. The only thing that can go wrong is that air bubbles may become trapped in the line, throwing off the reading. This tool is useful for determining whether one end of a wall is level with the other. It gives accurate readings over long distances. The water level can also be used in laying out a house.

There are other tools often substituted for the water level. A transit will do the job just as well. Transits are expensive, however, and not always dependable if rented. They take a degree of skill and practice to operate accurately. A simpler tool is the line level which is hung midway on a string stretched between the points to be compared. When using a line level one must be sure to pull the string as taut as possible. If it sags the reading will be off. A line level should not be relied on for accurate measurement. It may easily have an error of an inch in twenty feet.

Over shorter distances level and plumb can be measured using a two-foot or four-foot level. You'll find it convenient to own both sizes for masonry work. Most masons prefer a wooden one over a metal one because excess mortar is easier to clean off. A wooden level requires extra care and must be kept dry at all times. It is a good idea to oil and wax it occasionally. These precautions will keep it from warping. Using a level on irregular stone surfaces is not as easy as on wood or brick. You may have to take an average reading of the entire length. The accuracy of a reading over long distances can be increased by placing the level against a longer straight edge.

"average" plumb

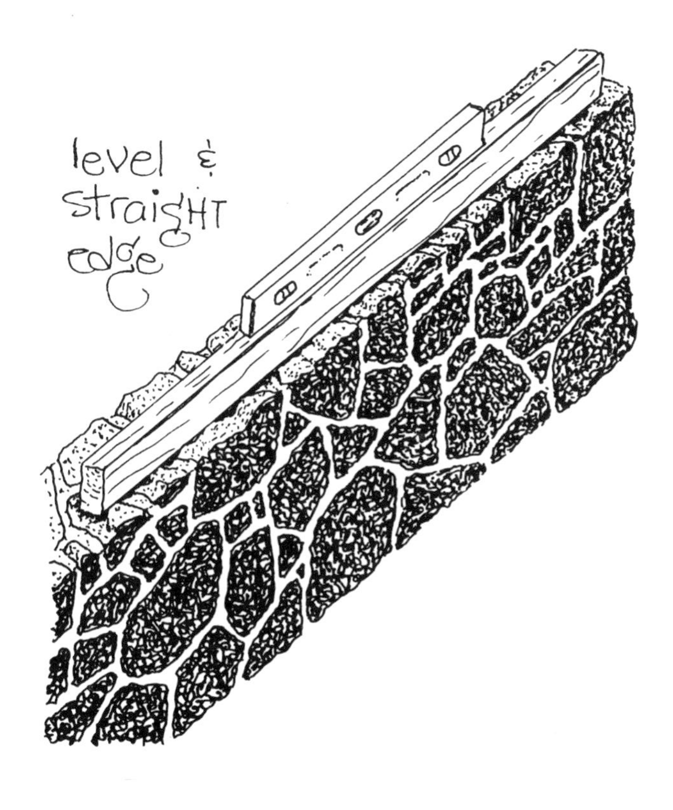

level & straight edge

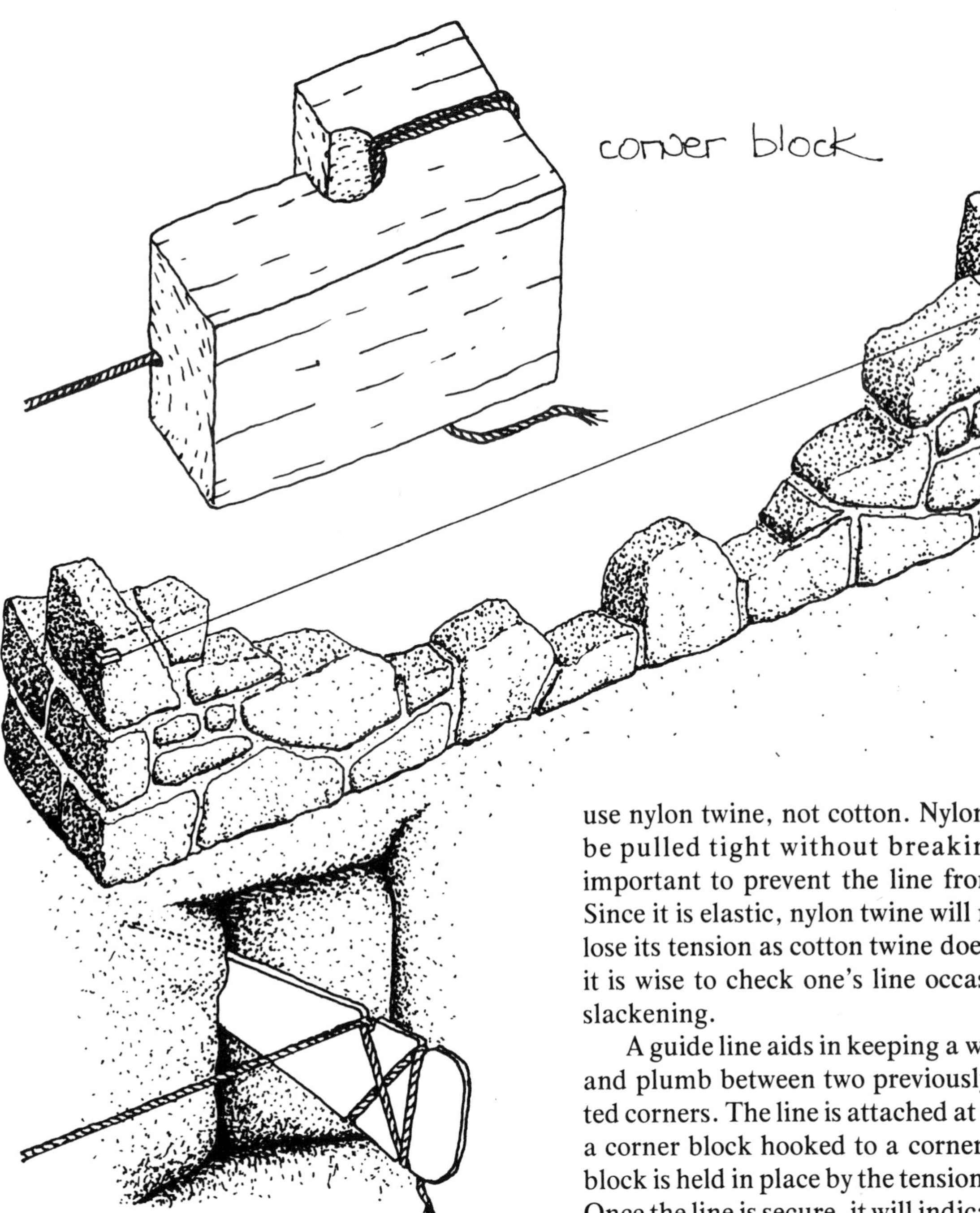

All the tools just mentioned aid the builder to determine whether a wall is going up plumb and level. It is, however, both time consuming and tedious to consult one of these tools every time you lay a stone in place. This will not be necessary since other aids have been developed which allow the builder to tell at a glance if the wall is true.

Most of these aids require the use of guide lines. When arranging guide lines be sure to use nylon twine, not cotton. Nylon twine can be pulled tight without breaking. This is important to prevent the line from sagging. Since it is elastic, nylon twine will not quickly lose its tension as cotton twine does. Even so, it is wise to check one's line occasionally for slackening.

A guide line aids in keeping a wall straight and plumb between two previously constructed corners. The line is attached at each end to a corner block hooked to a cornerstone. The block is held in place by the tension of the line. Once the line is secure, it will indicate whether the stone being positioned follows a straight course between both ends.

Corner blocks can often be found free of cost at masonry supply stores. They are designed for use with brick and block so they may not work well if your stone does not have sharp corners. In that case you may wish to fashion your own.

Corner blocks are of little use for interior walls where there are generally no corners to hook onto. For the inside of a wall a line pin is needed. A line pin is a small metal wedge. The pointed end is hammered into a joint between two stones near the corner. The other end has notches to which string can be attached. When these pins are positioned at each end of a wall they function like corner blocks.

More elaborate structures can be arranged using wooden framework to support both horizontal and vertical lines. If you choose to build one of these, make sure that the wooden frame is strong and well braced. It must not move if accidentally knocked.

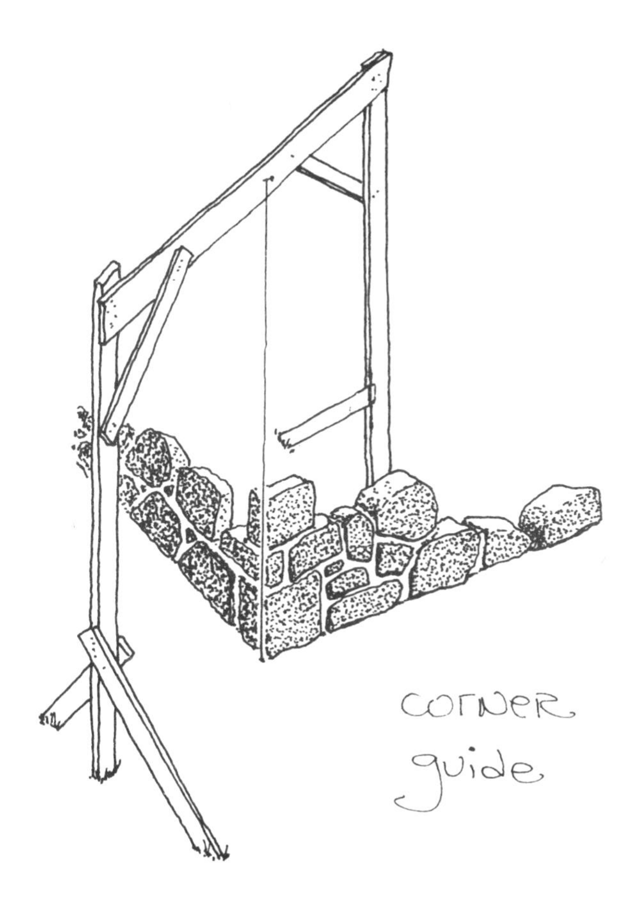

Useful guide lines are those defining the corners of the structure to be built. A level need not be consulted once you have erected a frame from which a plumb line hangs to indicate the corner. As the wall is built, the mason merely lines up cornerstones with the plumb line. These corner lines are also useful for keeping the wall between them straight. One can check whether a stone is in place by sighting across the two corner lines.

There are numerous ways to set up guide lines. They take planning but once built stone laying is easier and more accurate. In some cases an owner-builder may choose to build a curved stone wall. Curved walls are especially suited to stone masonry because, unlike wood or brick or block, nothing about stone prescribes its use in a straight line. If you do decide to build curved walls a different type of guide line must be used. It is more convenient to make the curve a segment of a circle. This makes the task easier to lay out and maintain the curve while building.

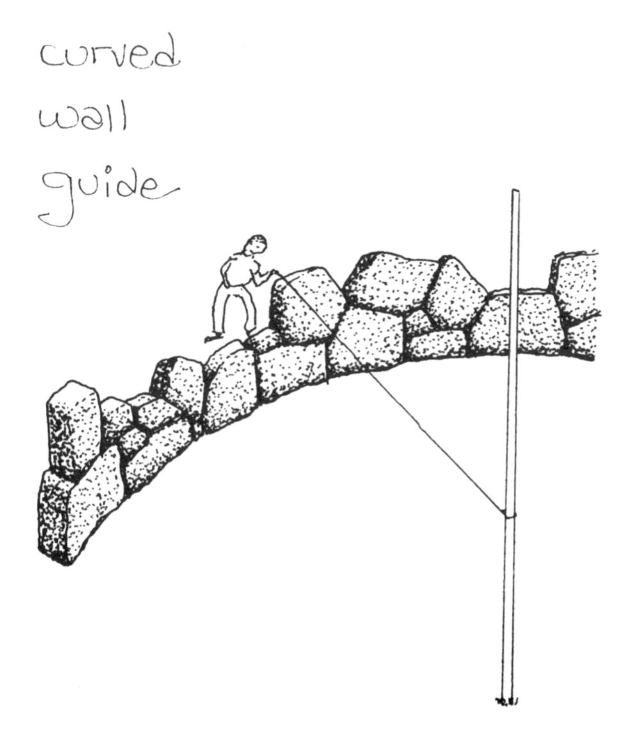

To create such a curve, first determine the center of the arc and erect a pole at this point. To this pole attach a wire whose length measures the same as radius of the desired arc. The end of this wire marks the face of the wall at every point it touches. Be sure to use wire, not twine, because you cannot measure exact distances with an elastic line. Remember that the end of the wire must be continually moved up the pole so that it is always level with the part of the wall you are building. If you do not, the wall will not be plumb.

Laying a Wall

Stone laying may be started after these important preparations. The basics of masonry were presented in the first section of the book; there are only a few bits of information to add which apply to the laid method. Although the wall will have two faces, it is important to think of it as a single unit and to build it that way. The two sides should be built simultaneously. Wherever possible, stones on one side of the wall should distribute their weight to the other side. Such an arrangement places stone with a wide bed back-to-back with narrower stone. On the next course the arrangement is reversed. In this manner joints are crossed within the wall as well as on the face of the wall. Use bond stones often; they span the entire thickness of the wall and present a face to both sides. Although bond stones are difficult to find and take forethought to position, they are necessary for unifying the wall.

When building a wall with a dead air space or insulation it is even more crucial to join the two sides at regular intervals. Bond stone can be used but a better alternative is to unite the sides with metal ties embedded in mortar. If insulation material like plastic or foam is used, the ties must pass through it.

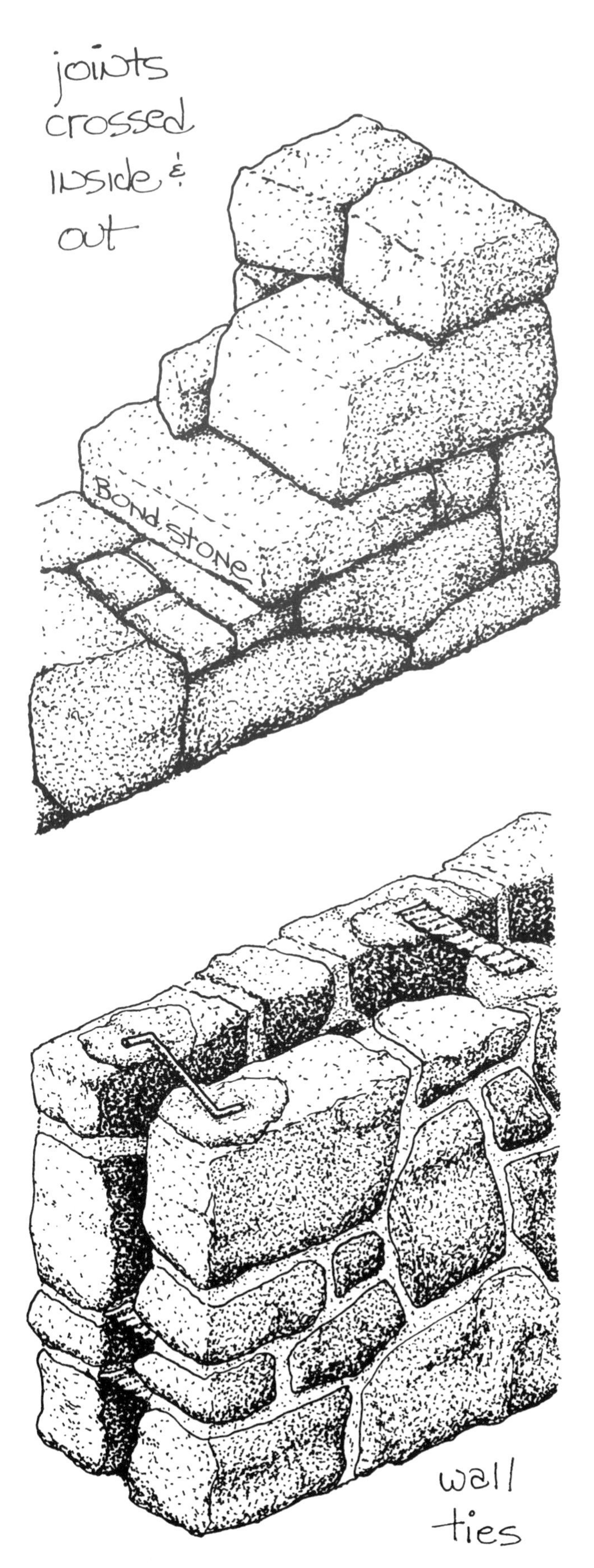

Reinforcing

Some masons prefer to strengthen walls with vertical and horizontal rebar (reinforcing rod) to insure that the wall is tied together, from bottom to top and side to side. It is a matter of personal judgement whether or not this additional precaution is necessary. In general it can be said that a single story stone wall seldom needs reinforcing. However, there are several conditions that make it desirable. In areas where earth tremors are a possibility, walls will need extra strength and continuity provided by steel rods. Also if you are uncertain of the stability of soil occupied by your house, a reinforced footing and wall may be needed. Moreover, walls subject to excess lateral strain from the house structure above them may require the added tensile strength of steel. In cases where reinforcing is advisable, the owner-builder might question whether or not stone is the best building material to use.

Openings

Where doors and windows are included in your wall, the way they are built depends largely on the type used. It is a good idea to buy or make windows before building the wall. Then, when it is time to position the window in the wall, brace it and lay stone around it. The frame should be secured to the masonry with partially driven nails whose heads are embedded in the mortar. Building the window into the wall is preferable to leaving an open space into which it is later fitted. In this way a tight, secure fit is assured. Door frames may be handled in the same way.

Windows and doors must have sills and stoops which slope toward the outside of the wall. This is necessary to keep water out of the house. If possible they should be made of one large, flat stone. Once you have built around windows and doors to their full height, it will be time to span them with stone. Wooden window frames cannot support the weight of stone. One common method places angle iron across the space to span the opening. A double-faced wall requires two pieces of stone placed back to back. If you have a number of large flat stone on hand you may be able to find a lintel stone long enough to span the distance without the aid of metal support. Such pieces are not always easy to find, especially when the opening may be three feet wide. One remaining alternative is to build an arch over the opening. Details for building an arch or eliminating the header will be given in following chapters.

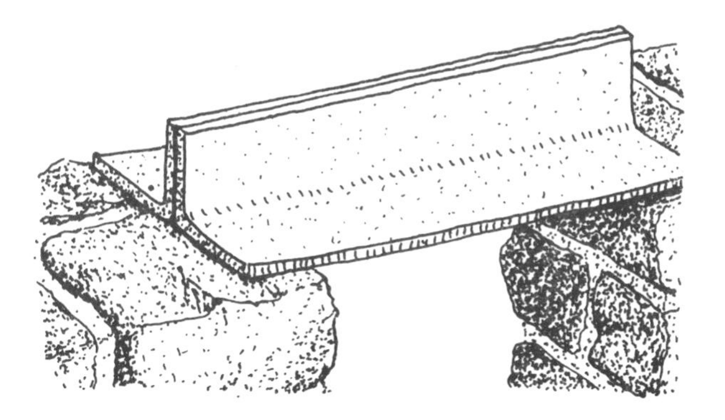

All the basic information has now been given, but the best way to learn to lay a stone wall is to follow a project from start to finish. One of the authors built his studio in the woods using the laid method. The following is an account of this project.

Studio in the Woods

The ideas presented in this section illustrate the procedure used to build a small studio in the woods. I have been working on this project over a period of years. Before starting I had very little experience with either stone masonry or carpentry. The laid method of stone building helped me to achieve certain visual effect while, at the same time, it enabled me to work in a slow, relaxed manner. I never attempted to run a race while laying up stone. To achieve the quality and control I desired, it was necessary to make an interesting arrangement with large and small shapes and to use a combination of different colored stones. Some of these came from a nearby river, others from a local dump where they had been used for landfill.

Frank Lloyd Wright and nature were my teachers for building with stone. I would often hike along rivers and creeks during summer months when the water was low, looking for rock outcroppings. I would climb close to study the individual character of stones and then would walk away to observe the total effect from a distance. Some of the rock in the outcroppings jutted further than others, creating interwoven shadows. The visual feeling was fantastic! However, in building it was not a case of imitating an effect of nature but of being inspired by the bold patterns that had been created. In no way do the final stone walls of the studio resemble the original strata although they suggest a similar feeling.

When I first thought about building a studio I was determined to use as many natural materials as possible. I looked for stone much as an old prospector would search for gold, wandering up and down rivers and creeks. After a period of time I was able to study the river to determine where the best stone could be found. At the end of summer when I had removed all the good building stone from a spot, I knew the spring floods would wash down new ones.

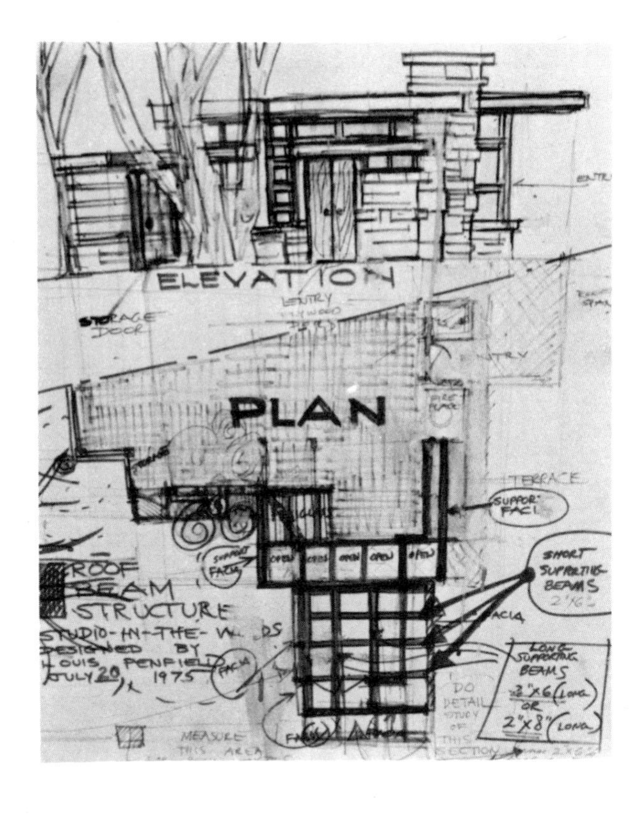

When I hauled stone to the site and unloaded it, I attempted to make separate piles of cornerstones, large flat stones and small stones. A single stack makes if difficult to pull any from the lower portion of the pile.

My function as an artist determined the basic plan of the studio. The site was located under the shade of a large, wild cherry tree where cool breezes pass during hot summer months. This location made an ideal place for working outside on August mornings and for relaxing inside during humid afternoons. I never sketched plans, at least not at first. I felt it was necessary to walk around the site many times to contemplate all aspects from all angles. Finally, I placed temporary stakes in the ground to indicate spaces for painting, sculpture and ceramics, each separate from the other. Lighting sources were approached in much the same way by placing markers to indicate where light might enter. To this day I do not know the measurements of any rooms within the studio. They were never thought of as numerical entities but as working spaces.

I'd never heard of anyone building a roof before walls, strange as it may seem, that is what I wanted to do. I wanted to work while it was raining without getting wet. It seemed plausible to put up some poles and build the roof even before starting the footing.

Holes were dug and partially filled with concrete. Posts were sunk in these holes and braced with long wooden members to make the posts stand straight while the concrete set up. In less than a week the roof was up and covered with permanent roofing. Thus, no time was lost during the spring, summer and autumn months due to inclement weather. Winter in a cold climate is a difficult time to build and although I enclosed the studio with plastic I did not attempt to lay stone when the temperature went below freezing. It is not necessary to build a roof first, as I did. With a bit of imagination a temporary covering can be constructed to waterproof the site.

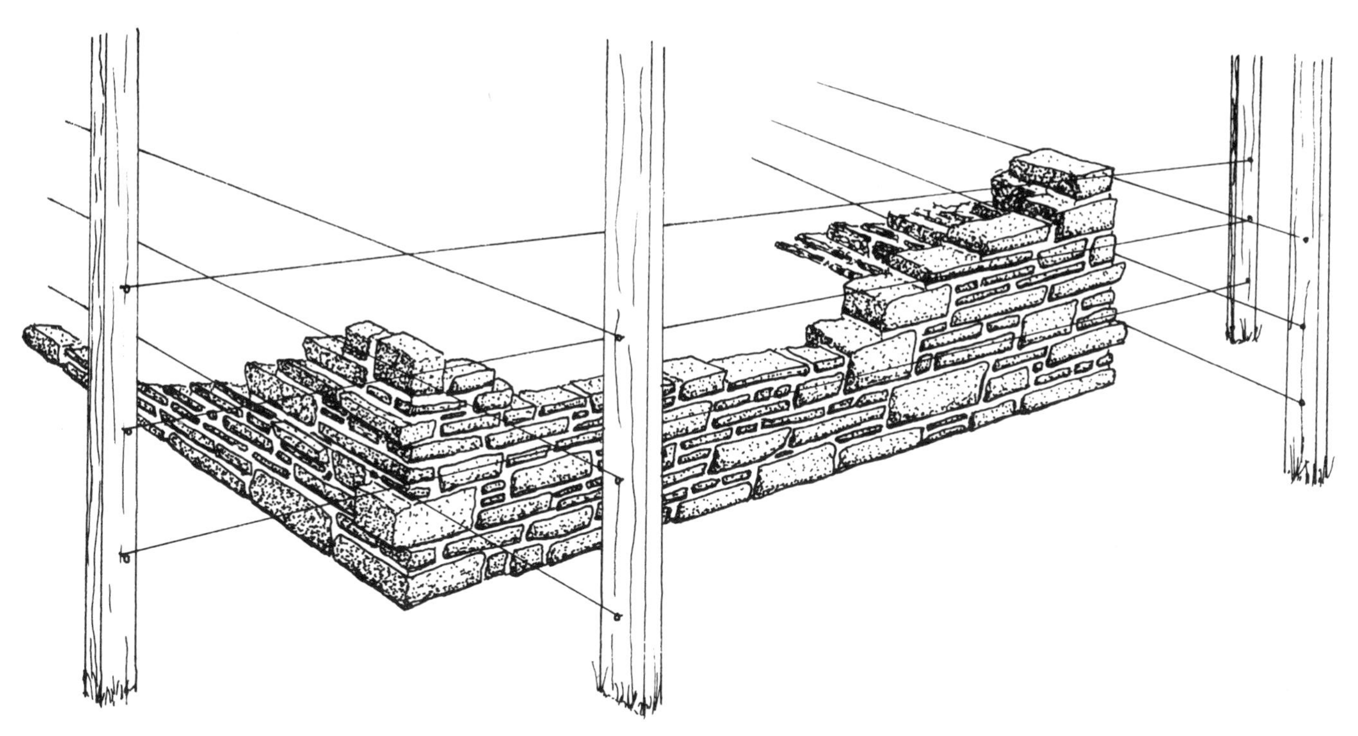

Next batter boards were placed around the outer portion of the building site, and exact right angles were established for all corners. Now I could begin digging the footing. I found myself using mostly the mattox and the pickaxe to loosen earth which was then shoveled out. This procedure allowed me to refrain from bending over with the shovel, preventing a sore back.

To fill the trench I made concrete from a mixture of water, cement, sand and gravel taken from the river bank. As the mixture was poured into the trench I also threw in small stone and culls. To get below the frost line my footing needed to be about forty-inches deep. I cut my reinforcing rod with a hack saw to fit the footing trenches, wiring it to the desired height. By using plenty of fill I was able to pour it all in one day.

With the footing poured I prepared to lay stone. First, I constructed wood and string guides to help me keep the stone plumb and level. Boards were driven in front of each corner as a true vertical guide for cornerstones. These boards were not placed directly on the corner but several inches back, leaving working room and enabling me to sight along the edge of the board to see if the cornerstones were plumb. A level can be used to keep corners plumb. However, in most of the walls I built, some stones were allowed to project beyond the face of the wall.

To keep the walls level and plumb I drove nails into the corner guides at sixteen-inch intervals, allowing at least an inch to protrude. I then ran twine from each nail to its counterpart on the opposite corner guide. With these lines I was able to see at a glance if my work was true.

I found it necessary to use two lines as guides for the width of the wall. Using the laid technique, both the inside and outside of the walls were stone. This was even more of a challenge because the whole construction was to remain hollow to create an insulating space.

With three or more strings mounted to the vertical boards, one directly above the other, I had an automatic guide to true alignment. When sighting down these strings they were seen as one line instead of three. I made sure outer edges lined up with them, except for the stones I pulled out to suggest projecting strata.

During the beginning stages of work I checked to see if the stone lay horizontal by kneeling down in front of the wall. If it did not, I tapped it down or lifted it up until it was level. This insured an effect of unity in the completed wall. After laying a few courses I would use the following approach for starting the day's work. First, I selected some stone I thought would fit together and then placed them on the wall without mortar. Starting this way I was able to go through the process of selecting, placing and changing stones around until the arrangement was pleasing. This was one of the most satisfying parts of the whole process. Once I decided on the final order, I removed the stones and placed them at the base of the wall in exactly the same sequence. Then I mortared each one in place.

After an entire course of stone was laid and partially set I raked mortar from the joints with a small trowel. This technique emphasizes each individual stone by creating strong shadows. When the mortar had set up for a period of time I brushed off the excess. I then put on rubber gloves and washed the stonework with a sponge and water to remove stains from the surface. When each wall section was completed, a mixture of water and muriatic acid was brushed on, scrubbed and rinsed.

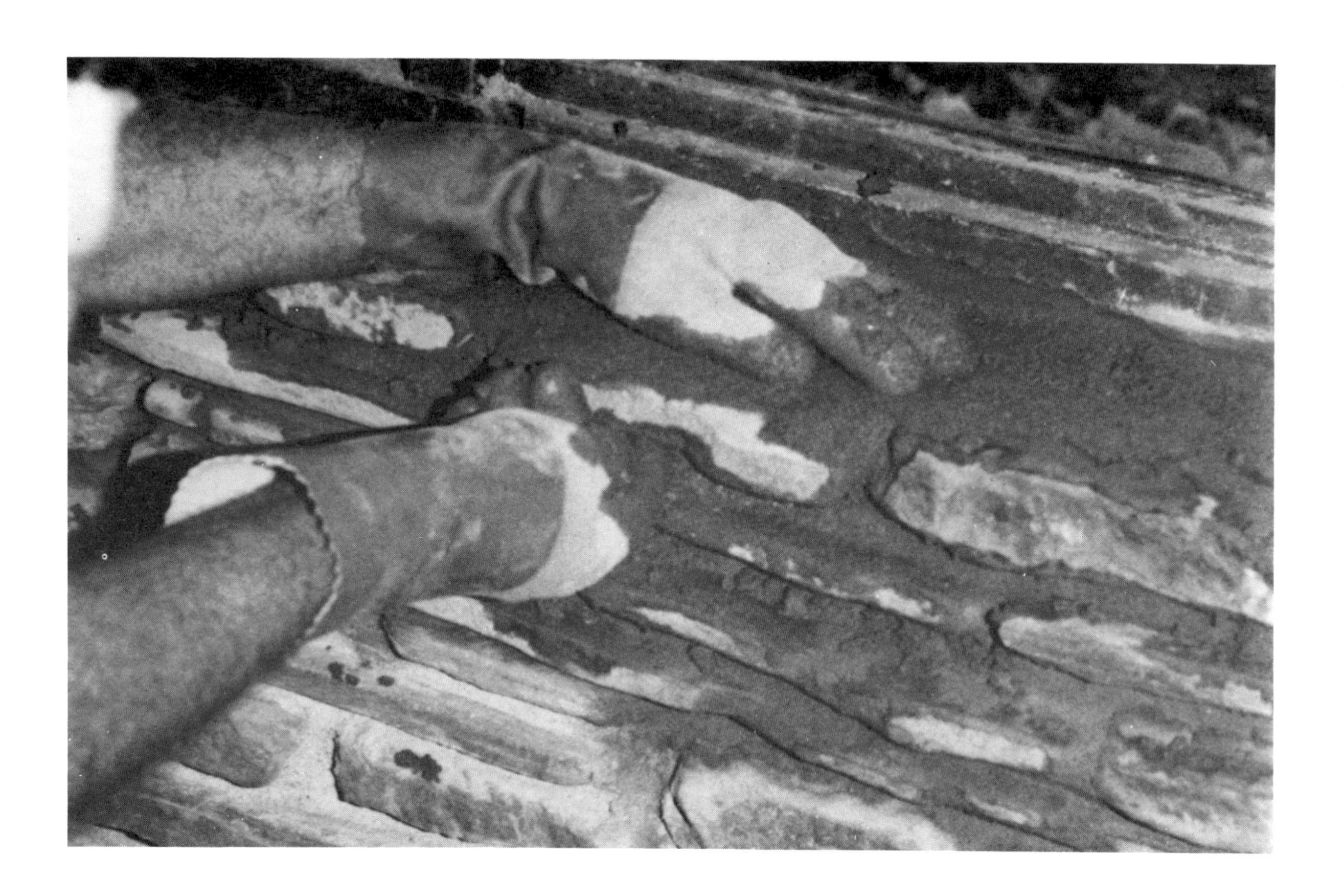

If you choose, you may begin work on your home immediately without initial experimentation or practice. However, I would suggest first trying to develop the feel of composing with stone. One of my original attempts was a cook-out area. I did not use a guide line but tried to combine uneven stones which were laid flat with some that were placed upright. The stone of these corners look jumbled and not carefully selected. I completed several walls and, although I enjoyed doing them, it took time to develop a method of laying that finally satisfied me.

My masonry work is limited to using river stone located near my studio and quarried stone that was hauled in. The flatness of these stone makes them ideal for laid masonry. This is a case where nature shaped and split them before they were washed up by spring floods. All the mason has to do is walk along, pick them up, carry them back to the site and lay them into a wall.

I am often asked when the studio in the woods will be finished. My usual reply is, "Never." When one section neared completion, I would plan an addition. For me it is an endless experiment.

Faced Masonry

Thus far we have shown stone masonry to be beautiful, durable, low-cost, maintenance-free and well suited to self-built construction. Despite all these advantages it is also labor intensive. A stick-built house may be erected in perhaps one-third the time and with one-third the energy. Fortunately, owner-builders generally have time and energy exceeding other resources.

Today it is customary in springtime for novice builders to move onto their land, living in a tent while they build. They expect to complete permanent shelter by winter. A solid stone structure, however, could scarcely be laid in one frost-free season. Due to this time pressure it has become popular to face exterior walls with stone. The skeleton of one's house can be erected using material with which it is faster and easier to build. Once the functional part of the house is completed an owner-builder can move in, leisurely facing outside walls with stone at another time — much as one might apply siding to exterior framed walls.

Facing a wall is faster and uses less material than solid masonry because only one side of the construction requires stonework. Stone facing is also easier to apply since there is backing against which to work. Besides these conveniences, insulation as well as plumbing and wiring is more easily installed in a faced wall than in solid masonry.

With this method one may build for the sheer love of working with stone since there is a reduction of the tedious work of building a monolithic wall. Faced masonry is generally more decorative than structural, so the owner-builder can give more attention to the design of the wall and can be more playful with stone forms. The pace for this part of the project may be more leisurely too, since progress on the rest of the house is not dependent on this aspect of the work. Still, when the job is done a permanent, strong, attractive and maintenance-free wall covering is the result.

Footings

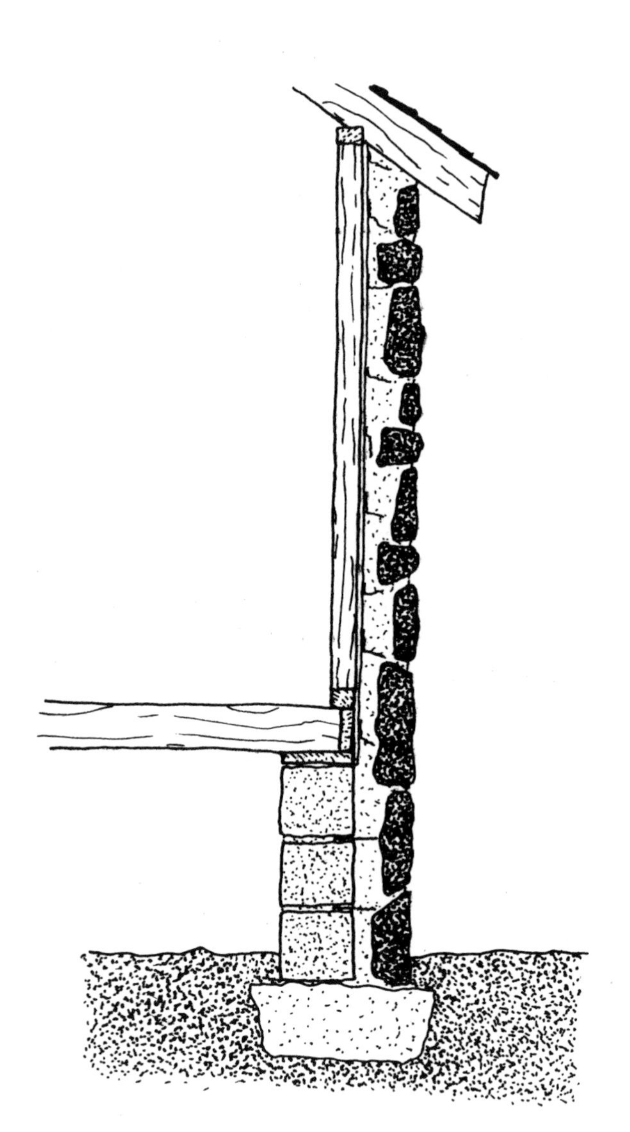

Although any wall may be faced with stone it must have a firm footing. Unlike wood siding, stone facing does not bend; it cracks. If a wall faced with stone begins to settle the facing will react in one of three ways. The facing will hold up the wall if it is well built with a solid footing; it may pull away from the settling wall and stand on its own; or it may itself be pulled down. In this case it will crack or crumble and fall. Often owner-builders encounter a situation like this because they decide to face walls with stone after their house has been completed. They must, therefore, dig a separate footing for the added facing. The walls and the facing, in this instance, are from bottom to top totally separate from each other and held together only by metal wall ties. Although this arrangement is often sufficient for their mutual support, the above mentioned problems may occur.

A surer way to build requires that one plan a project ahead of time and then pour a footing wide enough for both wall and facing. If they rise on the same footing they will more likely remain bonded together. If settling occurs at least they will go down together.

Backing

When facing, there must be some kind of firm, straight and plumb support (backing) against which to place stone and mortar. How thick the facing must be depends on the sturdiness of the backing. Stone facing of any thickness can be applied to any vertical surface. Some walls have a veneer of stone which is only an inch thick, while others have a facing that measures a foot in thickness.

When applying stone to a wall of poured concrete or cement block the facing is merely decorative. Such backing does not need additional support for its rigidity or strength. Aside from the decorative value, several other purposes are served by facing these self-supporting walls. Space may be provided between the wall and the facing for insulation or a vapor barrier. In the case of stud walls, stone facing may add rigidity and strength not attained by using wood alone.

It is possible to build a faced wall with minimal backing. A facing six to eight inches thick will need support only strong enough to hold itself up while the stone is being laid. Backing for such a wall may be made of two-by-twos braced in place and covered with builders' felt. Metal ties bind the two surfaces. The result is a finished wall with structural integrity.

As an alternative, temporary plywood backing may be secured in place while a facing as much as a foot thick is built against it. Metal ties are not required. Once mortar is set, backing can be removed and the wall will stand on its own. The newly exposed side may then be plastered. A variation on this method is achieved when one-inch-square wood furring strips are lightly tacked, horizontally or vertically, to the side of the plywood against which the masonry is laid. When the plywood form is removed the wood strips are embedded in the masonry. These strips are then used as nailers onto which inside paneling is later fastened.

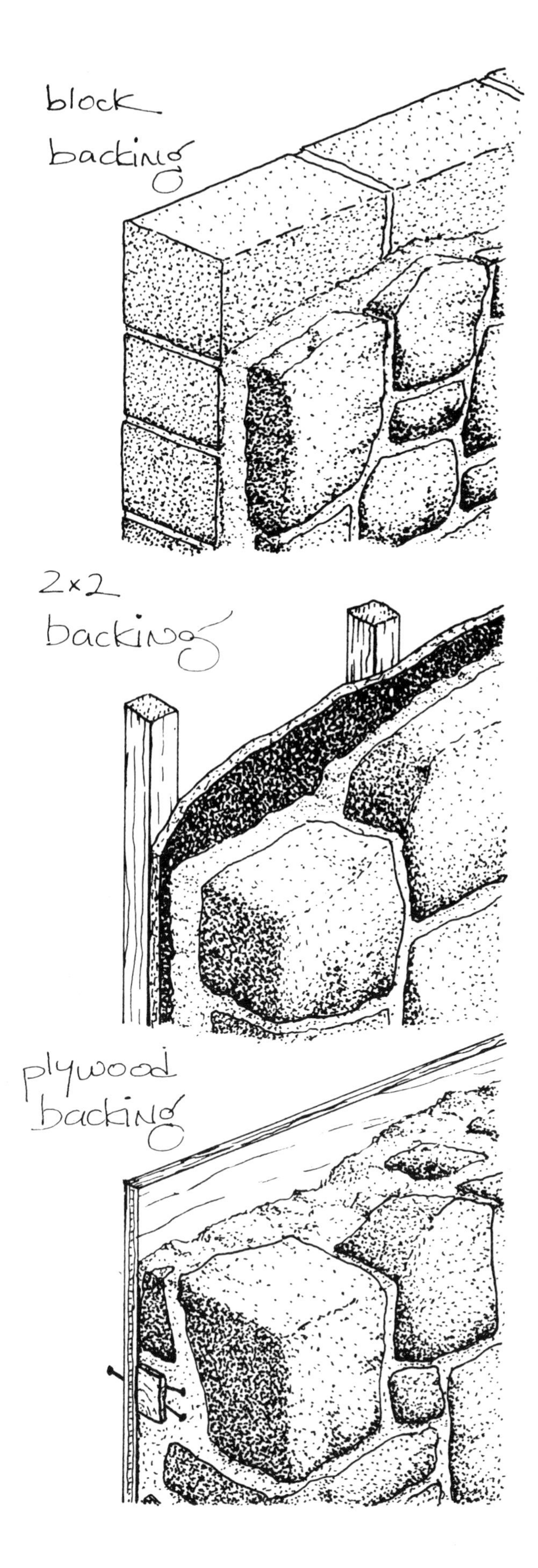

Preparing the Wall

Before a wall can be faced you must make sure it is secure and well braced. In the process stone is often laid so that it leans against this backing. Further pressure is exerted when wet mortar is filled between the stone and the backing which must support the facing until the mortar is set. If this wall is not properly braced it can bend, crack or fall due to the extra strain imposed upon it.

Before starting stonework you must choose insulation and decide how it will be installed. In the case of a frame wall it may be installed between the studs from the inside. However, if you plan to place insulation *between* facing and backing it must be done before any stone is laid. Popular materials used for this are sheets of polyurethane, styrofoam or insulating sheathing. Whichever is used, it must be attached firmly to the wall ahead of time. It is distracting to try to lay stone while simultaneously positioning sheets of insulation in or behind mortar and stone. Laying stone is a job in itself. 'Better keep it as simple as possible.

If the surface material covering the walls is not strong it can break, allowing mortar and even stone to fall through. Although one may face against plastic sheeting or builders' felt,

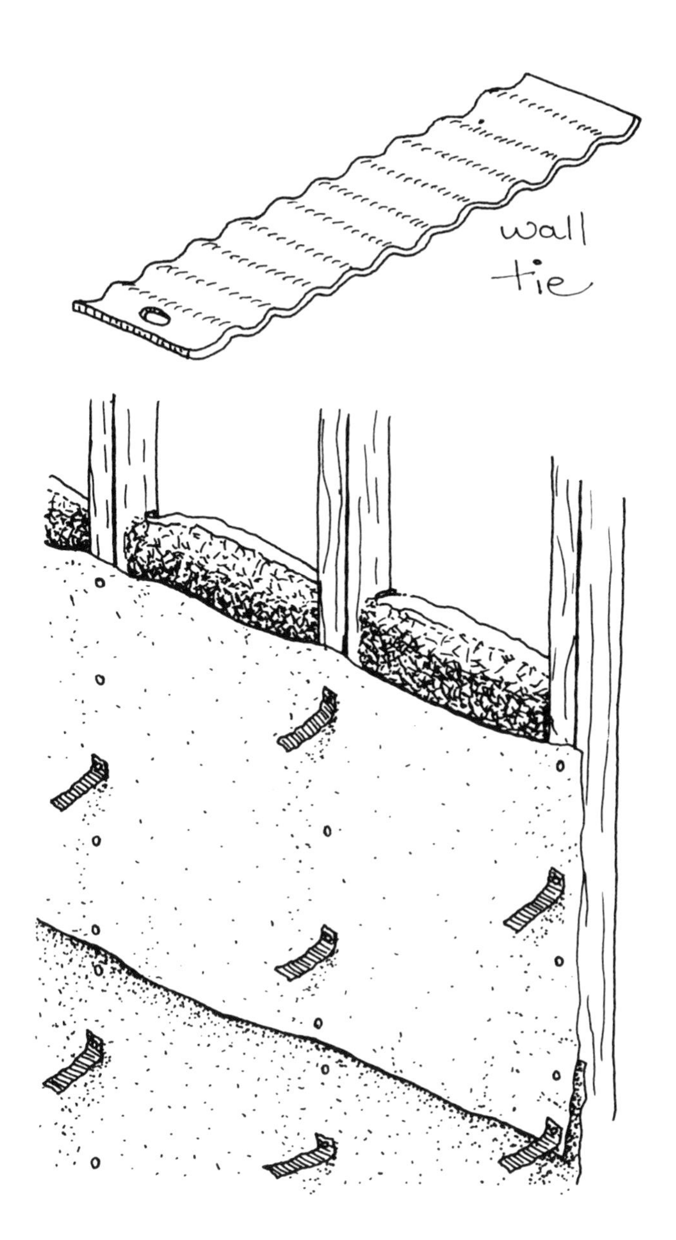

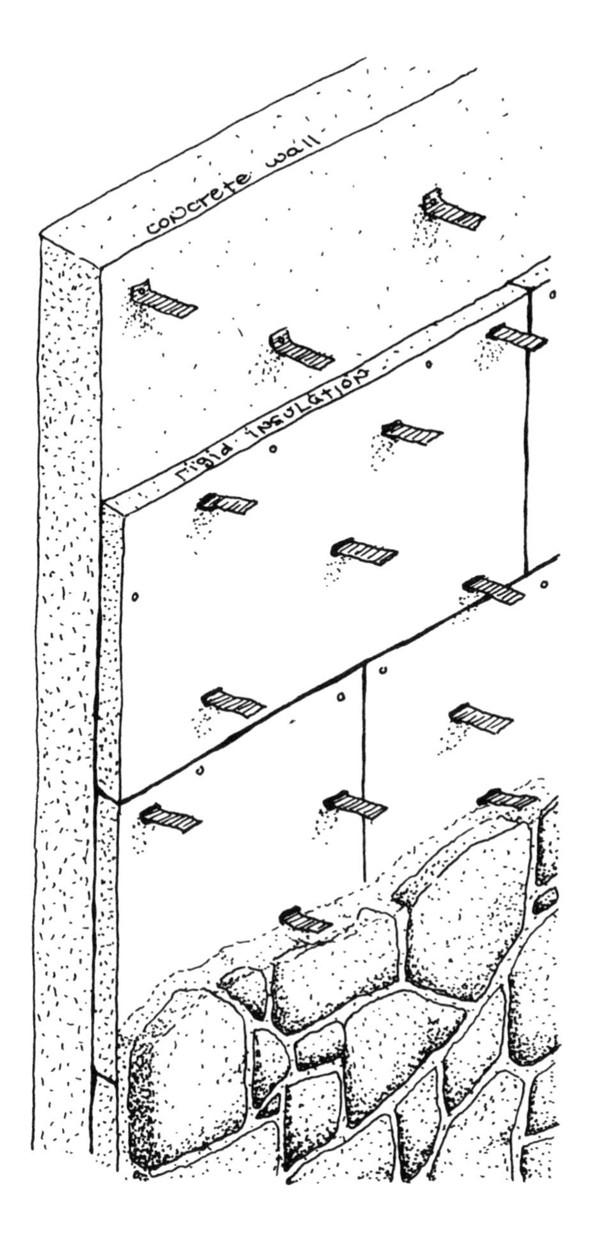

extreme care must be taken to fasten it to the framework so that it will withstand the pressure of wet mortar and stone.

Provision must be made for binding the stone face to the wall. To accomplish this, metal ties are placed either into or onto the wall at about two-foot intervals each way. To place ties into a block or concrete wall, they must be embedded when the cement is just laid or poured. If facing is to be added to previously laid masonry, it will be necessary to hammer ties onto the hardened wall with specially tempered nails. To fasten ties to a frame wall you simply nail one end to the studs. In all three instances the other end of the tie is embedded in the mortar joint of the facing. The two membranes are joined in this manner.

Other integrating devices are protruding bolts, nails, barbed wire loops or reinforcing steel projections. When building against block or concrete, bonding of the two surfaces is negligible. If insulating material is used between surfaces, it is especially important to have good binding devices that project through insulation to unite facing and wall.

Facing Stone

Before you choose stone decide how thick you want the facing. Another term for this is "bed width." If six-inch facing is planned it is pointless to collect thicker pieces unless some of them are to be allowed to protrude beyond the face of the wall. The size of the available stone may, infact, determine the thickness of the facing. If most pieces are six to eight-inches wide then it must be concluded that the bed width will have to be eight inches, whether or not a facing this thick is structurally required. If most available pieces are thin and flat you may have a choice. They could be stacked flatways for a wide facing or laid on edge to take advantage of the larger, more attractive faces.

Stone facing has to meet the same structural requirements as laid stonework. Although the faced wall may not be supportive as other members, it has to support itself. Pieces at the bottom of the wall must sustain the weight of several tons of stone and mortar above them. Therefore, make sure that all stones selected have a flat base and top — as well as a suitable face. This is essential when setting relatively thin veneer on edge. Quite often stone thought to be excellent veneer material breaks easily, creating thin, sharp edges. They are impossible to stack securely. Their blade-like base or top can sometimes be flattened with a stone hammer but make sure this is possible before gathering them.

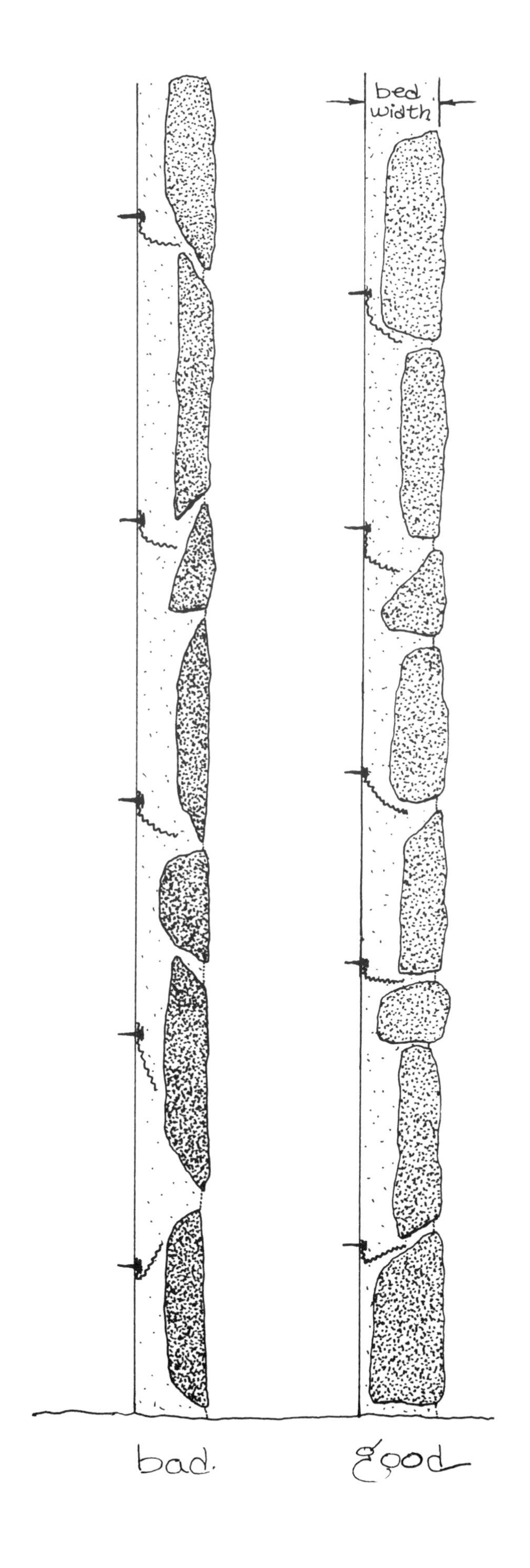
bed width
bad.
good

Facing a Wall

Laying stone for facing simplifies stonework by half, for the mason is concerned only with covering one side of a wall. Solid backing also makes it easier to lay stone. Another advantage of the facing method is that the mason does not have to set guide lines or consult a level to be reassured that the work is plumb. Merely measure from the face of the wall being covered to the face of the stone being laid and adjust the pieces to the desired width. Eyeballing (squinting across) from one corner of the facing to the other will reveal bulges or depressions. If this is done with every several stones laid, the wall will remain flat and plumb.

After each stone is mortared into place, additional mortar and rubble stone may be troweled into any remaining spaces. This backfill should be troweled until it mingles with mortar around it. Backfill can also be used to correct deficiencies in the bed of an irregular shaped stone.

Backfill may cause a problem when facing a wall, especially when laying thin veneer. Wet backfill may slump and push newly laid stone from position. Likewise, when tamping wet backfill into place this action may dislodge stonework. This is likely to happen when building quickly on a cool, damp day. A whole day's work has been known to fall because stone at the bottom of a freshly laid section is forced outward by the weight of stone and slow-drying mortar above.

Several precautions may be taken to prevent this occurance. The most obvious is to be sure that it is snuggled against the ones around it, held there by comparatively stiff mortar. The wetness of mortar determines how much it will slump. A drier, stiffer mix is harder to work but more stable; it holds its shape better when used to bed stone or backfill. Even then, if the bed width of a stone is thin it may easily be pushed out of place.

The entire length of the wall should be faced at once. Don't concentrate your attention on just one section. Lay a course of stone from one end of the wall to the other before backfilling any of it. This will give the stone a chance to set up in its mortar bed. Then when backfill is added it will not be as prone to push stones outward if it slumps.

The few places where a level or plumb line will be needed are at corners and around windows, doors and other openings. A simple way to indicate corners is to drop a plumb line at the top of the wall to the ground. This line will provide a constant limit toward which to build. When facing two adjoining walls, turning the corner is an obvious part of that process. But if you are facing a single wall you must think about finishing the edge where the facing ends. Details like this make a major difference in the final appearance of the work.

Curved Walls

Corner building is the most time-consuming aspect of the stone laying experience. Generally, the more inside and outside corners a building has, the more difficulty is experienced with the work and the more time and energy is expended. Musing over this for some time, one of the authors wondered if curved masonry walls could be built, making it unnecessary to contend with any corners whatsoever. The solution which evolved is so simple it is a wonder so few builders use curved masonry after it is demonstrated.

Any number of variations of curved wall construction are possible. Some are illustrated here. The procedure involves setting a radius pipe from which to swing a slipform. Concrete is packed into this form, composing a thin-shelled wall. Metal ties fasten the facing to this wall. This three-inch curvilinear wall may be erected with speed and ease, the building may be roofed and the interior inhabited before facing is applied. On one project, such walls sheltered a family for several years before they found time to insulate and veneer the outside of the building. This kind of priority construction makes good sense to shelter-needy families who are anxious to be free of the rent or mortgage-paying syndrome.

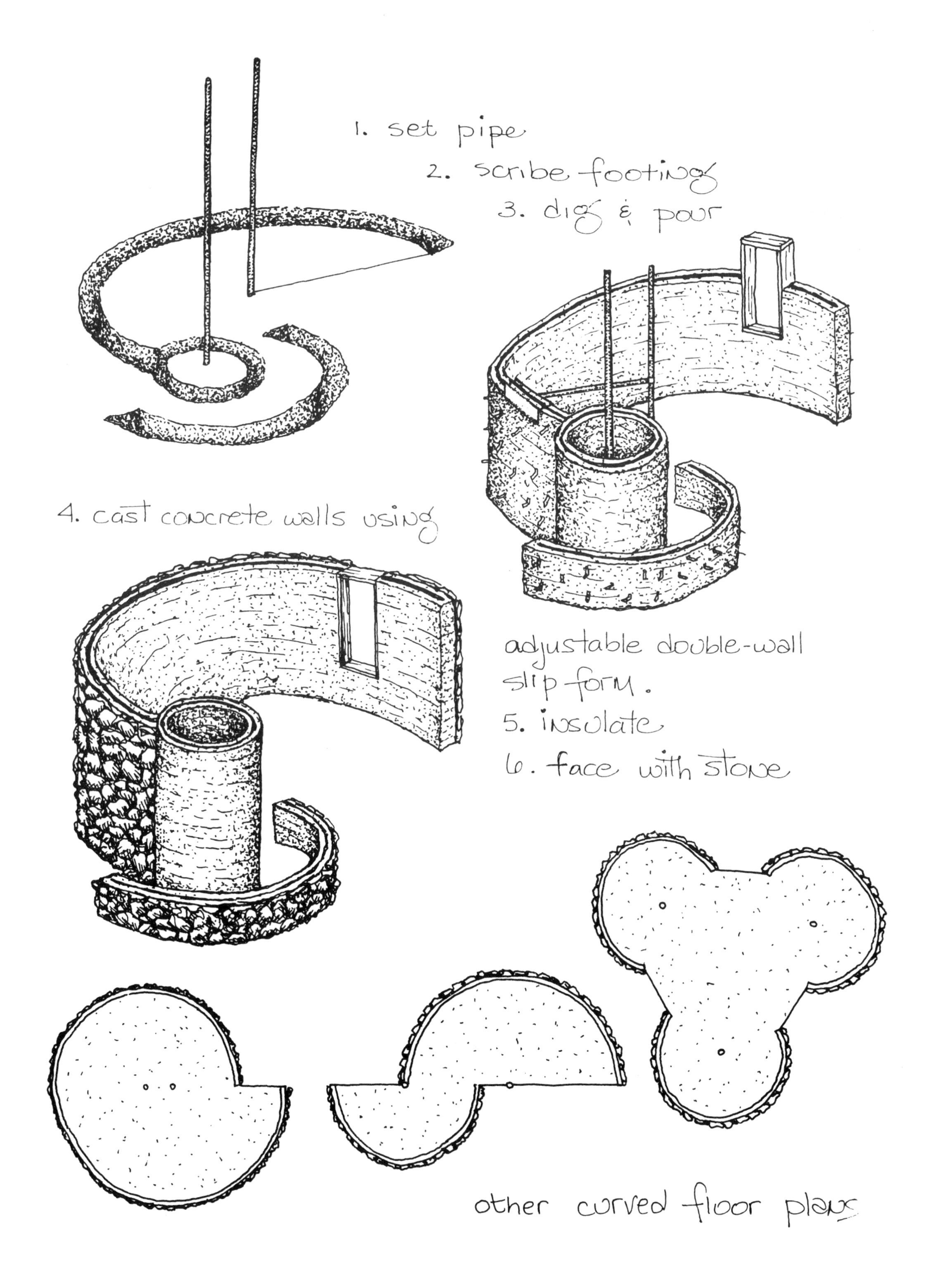
1. set pipe
2. scribe footing
3. dig & pour
4. cast concrete walls using
adjustable double-wall
slip form.
5. insulate
6. face with stone
other curved floor plans

Openings

Next to building corners, openings are the biggest hassel. A great deal of time and labor goes into facing around sills, headers and jambs, as detailed in the previous chapter. One alternative is to eliminate headers entirely. This is possible when the builder uses the type of construction where roof beams rest directly on walls, never over openings. Establishing the roof plate in line with the top of windows and doors eliminates laying stone over openings. Located at the highest section of the wall this is the area in which it is the most difficult to lay stone.

Combine as many functions as possible within each wall opening. One opening, for example, may serve as access to the outside while, at the same time, it provides interior light and ventilation. Another good practice groups and centralizes openings, preventing the punched-hole effect which characteristically mars much contemporary building design.

The facing method is a relatively recent innovation. Purists may not, for this reason, wish to use it. To be sure, the method lacks the old-timey appeal of a substantial, laid stone wall. There is for many of us, however, another resource which is becoming more scarce — time. The convenience and relative ease of erecting stone facing allows for the optimal use of this resource.

Formed Masonry

The method of packing stone rubble and cement into a movable wooden trough to form walls probably originated in this country over a century ago. In the 1840s a man named Goodrich invented a movable wooden form for casting stone walls. Orson Fowler mentioned Goodrich's system in his book, *The Octagon House,* published in 1848, and he went on to perfect Goodrich's method. Fowler considered his octagon-shaped stone house to be an acceptable solution to peoples' housing needs.

Fowler's unbounded enthusiasm for formed masonry wall construction failed to survive his lifetime. Fifty years passed after his death before a revival of interest in formed masonry construction occurred. Eventually, the eminent New York architect, Ernest Flagg, perfected the design and construction of small, low cost housing cast in stone.

Flagg

Flagg carried on a successful architectural design practice for over fifty years. His enduring interest was in small house building, although he designed a number of major structures like the Singer building in New York City — in 1908, the tallest commercial structure in this country — the United States Naval Academy at Anapolis and the Washington State capitol in Olympia. In 1921 Flagg published two hundred pages of home buildings designs and techniques in his book, *Small Homes, Their Economic Design and Construction*. (Charles Scribners' Publishers, NY)

About the time Flagg started his practice in the early 1890s, he had occasion to build a rubble masonry garden wall at his Long Island home. Observing the slow, tedious nature of laying stonework, Flagg was prompted to experiment with a "false work" of planks and uprights to form the wall. He first placed uprights at either side of the wall to be built. These were braced to stakes in the ground a short distance away. Planks were then placed between the uprights and rubble was poured between the planks. Notwithstanding the time and trouble required to build the form, Flagg found that the time spent laying his garden wall was decreased by fifty percent. Flagg used steel pins, not nails, to anchor the planks to the uprights. The planks could then later be reused simply by removing the pins and moving the planks upward.

Flagg worked on what he called his "mosaic rubble" wall-building system for the next twenty-five years. He built countless stone houses and made improvements in his forming method with each house. Perhaps the biggest obstacle encountered was the removal of the planks after the stonework had set up. The planks could not be removed until the uprights were withdrawn. It proved to be a chore to remove the uprights because the diagonal members bracing them were in the way. Flagg finally solved this problem by

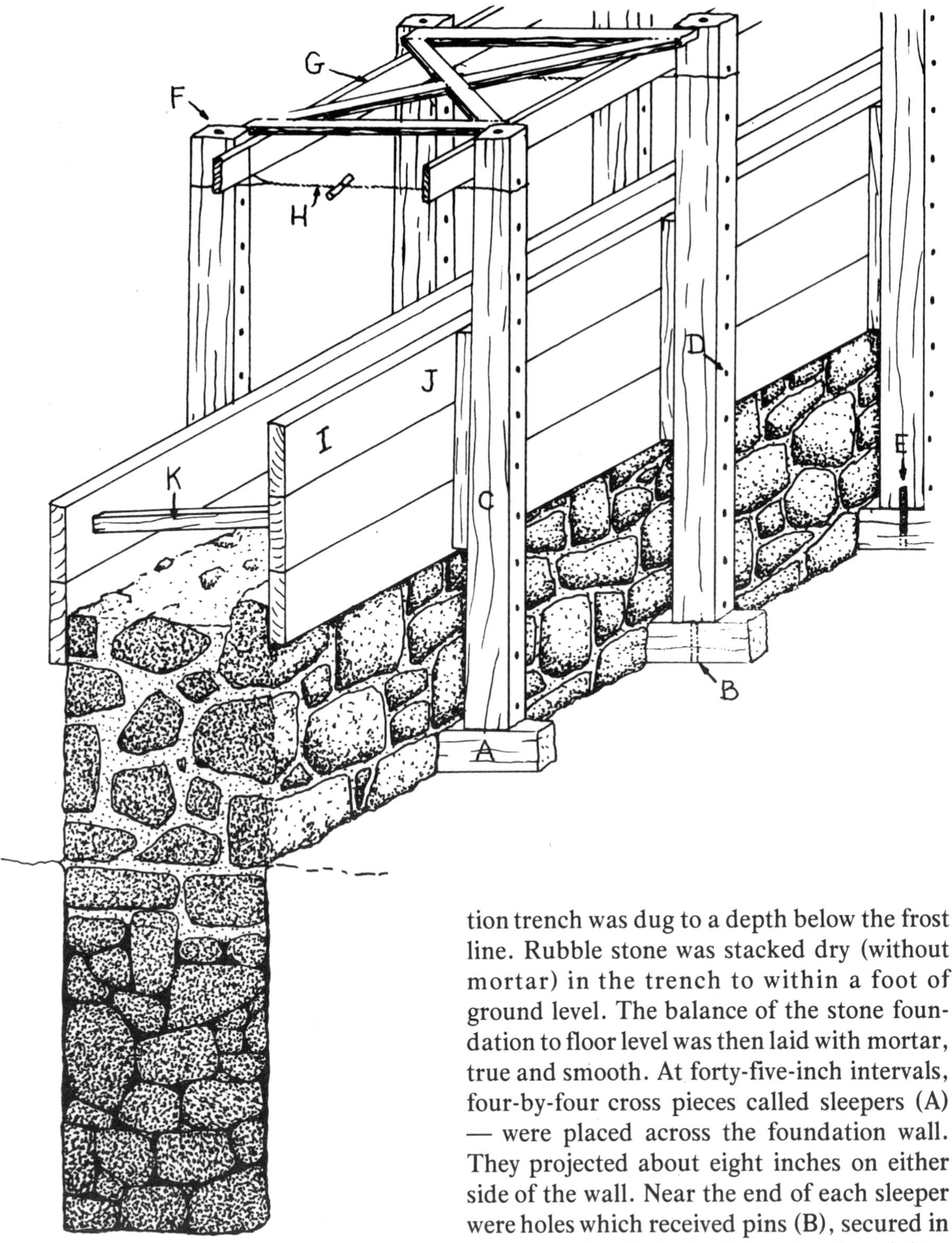

anchoring the uprights onto a system of sleepers which were integral with the wall itself. The following is a brief description of Flagg's mosaic rubble construction.

Around the building perimeter, a foundation trench was dug to a depth below the frost line. Rubble stone was stacked dry (without mortar) in the trench to within a foot of ground level. The balance of the stone foundation to floor level was then laid with mortar, true and smooth. At forty-five-inch intervals, four-by-four cross pieces called sleepers (A) — were placed across the foundation wall. They projected about eight inches on either side of the wall. Near the end of each sleeper were holes which received pins (B), secured in corresponding holes in the ends of uprights. The sleepers were tapered and greased before use to facilitate their removal when the wall was completed.

Each four-by-four upright (C) had holes drilled at six-inch intervals (D). At one side of

its central axis, a projecting pin (E) was located at one end of the upright and a corresponding hole (F) was found at the other end of the upright. The top of the upright was held against an alignment truss (G) by wires (H) looped around each pair of uprights. These wires were twisted to make them taut. Planks of two-by-tens (I) did not come into direct contact with the upright but were separated from them by short pieces (J), called release sticks. They were prevented from falling inward by other short pieces (K), called spreaders. Flagg found that only two or three planks could be used in one forming sequence until the mortar was set. No mortar was used between stones at the face of the wall. Pieces were placed against the planks like pieces in a mosaic and concrete was shoveled behind them.

Flagg's fifteen-inch thick walls consisted of a facing of stone and a backing of rubble and concrete — nothing more. He would cast his wall two-feet high in one day. The following morning, the pins were released and the planks were removed and raised to the next height. Few planks were needed because, as cement set, lower ones could be removed and used at a higher level.

To finish his mosaic rubble walls, Flagg insisted on pointing them. This he did by placing cement on a mortar board, holding it against the wall and shoving the mortar into joints with a pointing trowel until they were completely filled.

Flagg devised an ingenious method for doubling the use of the uprights as support for scaffolding and runways. Wherever possible, Flagg used ramps so that cement and stone could be wheeled, not hoisted, to the height of the form. More will be said later in this section about flagg's design criterria for building low walls.

Thousands of houses were formed of stone following the publication of Flagg's *Small Homes* book, which fortunately appeared prior to the beginning of the Great Depression. The School of Living nationally popularized the Flagg building system with its "How to Economize" publications. A number of Flagg-built homes were erected at the original site of the School of Living in Rockland County, New York. One of the models, the 2200-square-foot house pictured below, cost $4,000 to build in 1938.

As more and more builders were attracted to Flagg's ideas variations and modifications of his forming methods appeared. In some cases, his original systems were no longer recognizable as people built stone houses and wrote about their experiences.

Peters

Like Flagg, Frazier Peters was an architect-turned-builder who was enthused about prospects for building stone-formed low-cost houses. In his two books *Houses of Stone* (1933) and *Pour Yourself a House* (1949), Peters developed forming techniques where Flagg left off. He recognized one major drawback involved with stone cast walls — the additional expense of erecting a separate, wooden frame wall against the masonry in the house interior. Essentially, this procedure, called furring is necessary to insulate and moisture-proof a wall.

Each of the builders described in this chapter approached wall furring differently and each achieved varying degrees of success and economy. Some builders nailed or glued furring strips onto the finished masonry. Other builders set furring strips *into* the masonry to serve as lightweight nailers for a second row of strips applied after the forms were removed. In both cases the main detriment of using furring strips is that, in time, they gradually loosen and tend to show dry rot. And invariably the stone wall is irregularly aligned, making it difficult to set nailing strips either

against or into the masonry. Stud framing was nailed to the furring strips to finish inside walls. We authors feel that this practice constitutes a waste of material and labor and a misuse of stone and concrete.

Peters had the foresight to use insulating sheathing for weather proofing on the inside of the form. Bolts through this sheathing held it permanently to the finished wall after forms were removed. This result is close to the solution we recommend. Peters used celotex board, but, no doubt would have used urethane insulation board for its superior insulative value had it been available. Lifestyles have also changed since the 30s and people today are more accepting and even prefer a wall finish that has the rough hewn quality of stone. Logically, the place for insulation board is in the *middle* of the wall, not against the inside face.

Peters refined some noteworthy procedures for forming stone walls. He, first, amassed enough material in the center of the project to complete it. Then a trench was dug and a perimeter foundation was poured. Uprights of two-by-fours were built completely around inside and outside walls on two-foot centers. Slatted sheathing form boards of one-by-sixs were nailed to the outside uprights. Along with the inside uprights before any stone was laid, the entire outside form for the house was erected. As wall pouring commenced additional sheets of twenty-four-inch insulation board were placed against inside uprights. It was convenient to build interior walls with stone from the centrally-placed stockpile.

Nearing

Possibly the most far-reaching influence for building with forms comes from Helen and Scott Nearing's book *Living the Good Life,* (Social Science Institute, Harborside ME 04642). Since the early 30s the Nearing's have built more than a dozen stone structures on their successive homesteads in Vermont and Maine.

The type of form work employed by the Nearings was, again, an adaptation of the Flagg method. They built forms using three pieces of one-by-six boards anchored to a framework of two-by-threes. The unique feature of the Nearing's method was their system of hand-over-hand forming. After one set of forms was poured, an upper set of forms was tacked to the lower ones and poured. The lower set was removed only after the upper set was firm.

The drawback of this system is the expense of investing in a double set of forms extending around the entire perimeter of the building. One builder calculated that the cost for the material to build a double set of forms for 1500 square feet of housing came to $400. Only if the cost of each set of forms could be spread over a number of constructions would the cost be reasonable.

Corey

Paul Corey cast his first stone house in 1930. This project and the two houses that were to follow led to the publication in 1950 of his book, *Hand Made Homes* (William Sloan Publishers). His form construction method was simple, and eventually it evolved some significant improvements. We will briefly describe each project to show how they developed.

On his first house, Cory used two-by-fours for upright supports which spanned the height of the wall. The uprights were spaced four feet apart, and one-by-eight shiplap

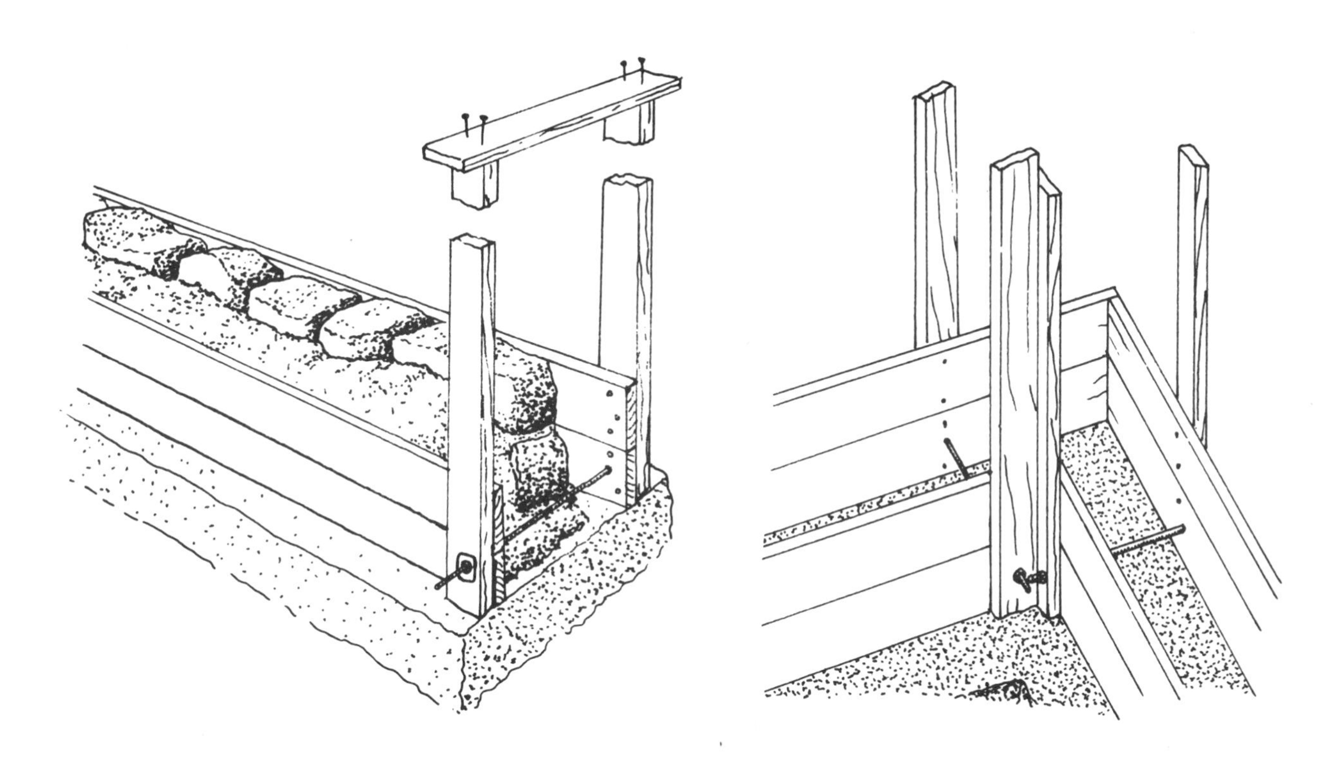

siding was nailed against the inside uprights. On the outside he spaced two-by-four uprights to match. Near the bottom of the upright he drilled a five-eights-inch hole through both frames and secured eighteen-inch long half-inch bolts.

At the corners, Corey placed the uprights at right angles so that the bolts would securely tie the two members together, as illustrated in the accompanying sketch. As stone and cement were placed together in the trough, additional boards were added. The wall progressed one board height at a time.

As soon as the cement in the final course was set, Corey loosened the boards tying the uprights together and pried the form off the bolts. The unveiling of the wall presented a straight line of stone on the outside and a smooth cement surface inside. When this method of stone wall construction is used only the first two boards need to be nailed to the uprights. Additional boards can be set in on top of the lower boards and propped until the rock and cement hold them in place.

No reinforcing steel was used in Corey's first project — nor did he allow for an insulated air space. The inside of the foot-thick stone wall was merely painted. The Corey family lived in this Putnam County, NY house for ten years and never regretted not having air space insulation. Furring strips, however, could have been easily anchored to the masonry wall using Tuff-Bond construction adhesive and Gemco anchor nails. Perforated plates with sharp pins extending from centers are stuck to the walls with adhesive, furring strips are driven onto the pins and the protruding ends are bent over. Gypsum board or wood paneling can then be hung on the furring strips.

Corey's second house included a three-fourths-inch dead air space between the wood and concrete. For the inside form he nailed one-by-eight shiplap siding to the narrow edge of rough cut one-by-threes which were located on sixteen-inch centers. These one-by-threes were set in the wall and were as long as the wall is high. The frame was plumbed and held in position using large stones at the bottom and tie braces at the top. The outside form consisted of one-by-eight shiplap siding nailed to two-by-fours at four-foot intervals. The bottom ends of these two-by-fours were held in position with large stones or stakes.

After laying a course of stone against the outside form, cement and smaller stone were used to fill the space against the inside form and around the one-by-threes. Corey drove twenty-penny nails through the one-by-threes on two-foot centers to embed them in the concrete, preventing them from loosening once the wood shrank.

For this type of form-built stone wall one needs only four or five form boards. A second board is set on top of the first on the outside and on the inside. Double headed form nails are used to nail the boards on the inside. After the second course of stone and cement has set up, the board below is moved up to a new position above. On the outside, the second board is slid from behind the two-by-fours and placed higher for the following course. Corey found that the outside form for this type of construction does not have to be strong. It must only be strong enough to keep the outside surface straight.

Corey preferred to work alone when he built stone walls. He accumulated the stone, mixed the cement and laid up the stone himself. This way he was familiar with each stone in the supply pile. When he came to a place in the wall that required a stone of a specific shape, he knew exactly where it might be found. This saved him a great deal of time. If more than one person is to work on a wall, Corey suggests that others accumulate their own personal stockpile of stone and work on their own section of the wall.

This type of formwork is not designed to take the pressure of large masses of unset cement and stone. One should build horizontally rather than vertically. No more than two one-by-eight heights should be laid around the entire perimeter of the house in one day. In the morning following a day's pour, the lower boards can be moved up and work may be continued. Corey also found that it was easier to work from the level slab inside the house.

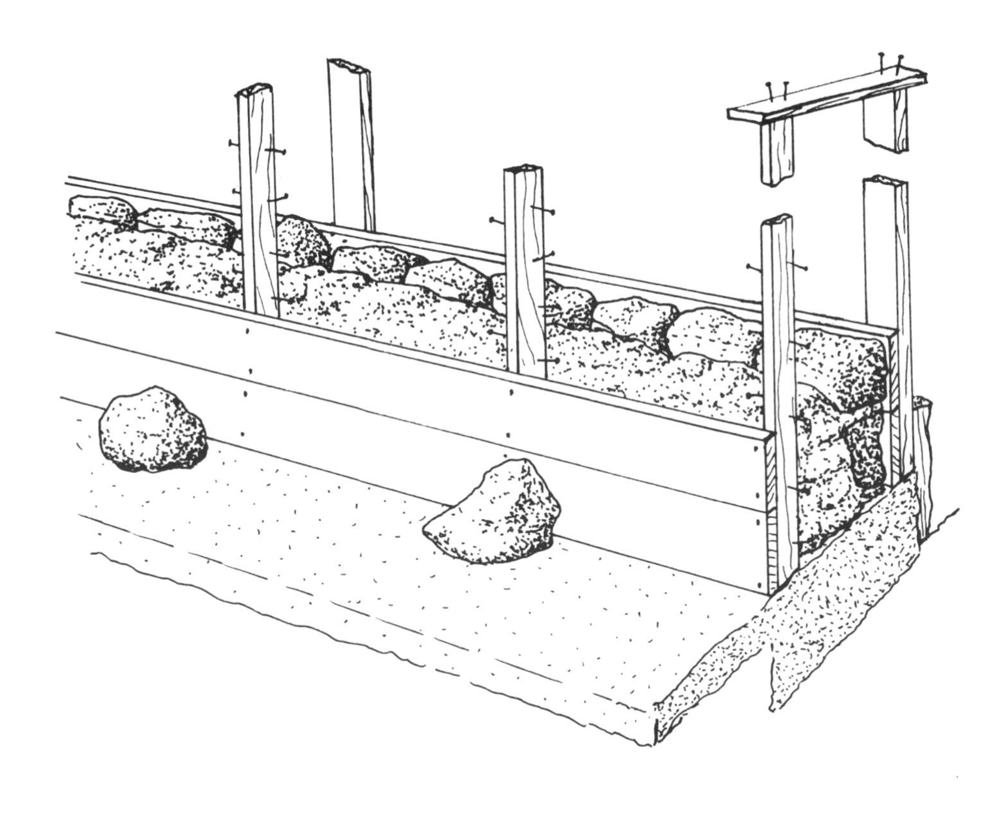

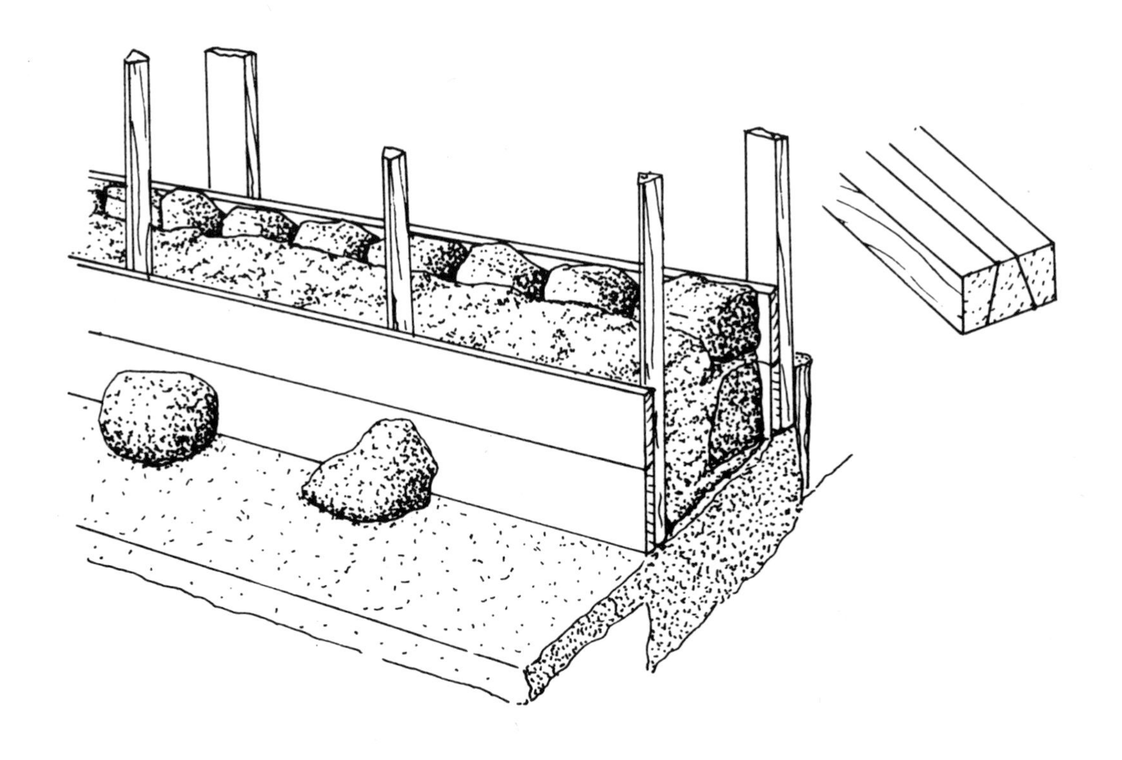

The third house Corey built was located in California, and the forming system he used on this house was similar to that used on his second house. He did, however, add steel reinforcing rods to the fourteen-inch thick walls. Instead of using one-by-threes for nailing strips, Corey took advantage of power tools, not available when he built his first two houses, and he ripped two-by-fours. They were cut into three wedge-shaped pieces which were set with the wide base inside the wall at sixteen-inch centers, as illustrated in the accompanying sketch. After the wall was completed, he furred it with one-by-twos nailed flat to the three-fourths-inch exposed edges and hung gypsum board on this frame. All the wood members remaining within the wall were treated with a wood preservative.

Corey also had the use of a cement mixer on his third house. He found that a regular cement mixer was best to fill the space between the outer stone wall and the inner surface.

The larger aggregate of this mix did not seem to affect the strength of the wall and was considerably cheaper when compared in cost with a mix of one-to-five cement and sand.

Everyone who has written about building a stone-formed house appears to have selected their own particular technique. For instance, Flagg suggested a ratio of 1:5:10 mortar mix (concrete) of portland cement and sand and pea gravel. Peters prefers a mix of 1:3:4 and Nearings use a mix of 1:3:6. For general form work we feel that the Peters proportion is preferable. Actually, water content is more important than mix proportion. A "sticky" cement mix, consisting of not over five gallons of water for each sack of cement, should be used. Concrete should be packed, not poured, into the form. About two-thirds of the space in the form should be filled with stone — the rest is concrete. A stone wall cast in forms can be built twice as fast as a laid masonry wall. Corey claimed that a builder working alone, mixing cement, carrying stone and mortar and laying up wall could average forty square-feet a day.

Watson

Lewis and Sharon Watson built their stone house using only one set of forms, merely anchoring the forms to vertical two-by-two uprights. Tie wires were looped over each pair of uprights and were tied through the form to support it prior to pouring. The Watsons constructed sufficient forms to build from one doorway or window opening to the next. This sequence involved one-half of the actual perimeter of the building. An entire floor-to-ceiling panel was poured before moving on to the next segment. A delightful book describing their homebuilding experience has been written by the Watsons. *How to Build a Low Cost House of Stone* is available from them for $3 at Sweet, ID 83670.

Flagg's ridge dormer

Fryer

The desert home of Richard and Sandi Fryer was designed by one of the authors in an attempt to minimize form building. Every wall in this hexagonally-shaped house is an identical fifteen feet in length. Thus, only one pair of fifteen-foot long forms was needed to build this entire house. Furthermore, in order to eliminate corners a simple buttress form was used to connect walls at their juncture.

The outside walls of the Fryer house were kept at minimal height. Obviously, high masonry walls are considerably more expensive to construct than low ones. This is the primary lesson learned from the work of Ernest Flagg. The lateral walls of a single story Flagg house were only five feet high. The low height of the wall was compensated by the construction of an ingenious ridge dormer, a feature found in almost every Flagg house. Again, by ending the top of all door and window openings at plate line, wall heights are reduced, eliminating a lintel or arch to span these openings.

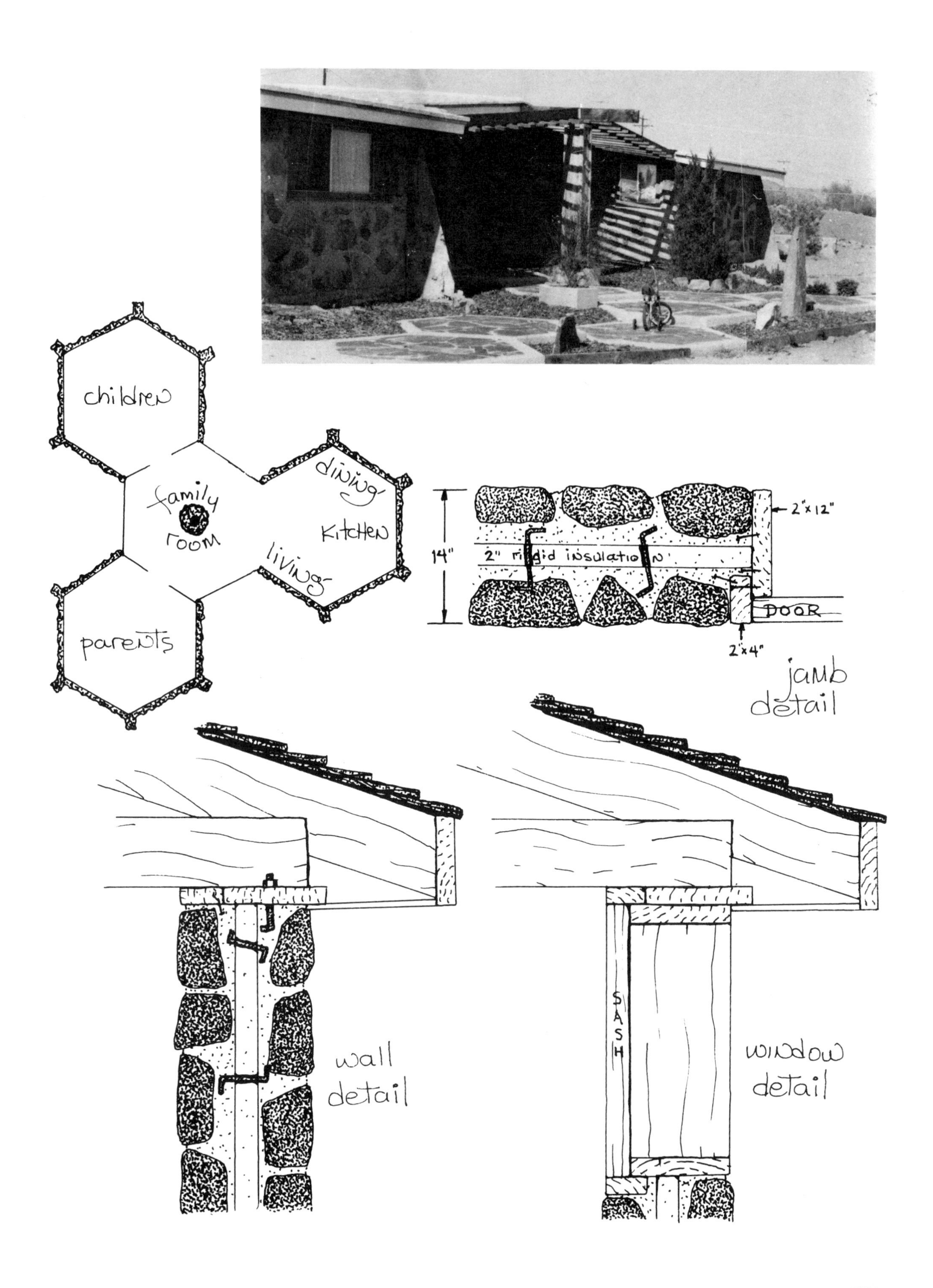
children
family
room
dining
kitchen
living
parents
14"
2" rigid insulation
2"x12"
DOOR
2"x4"
jamb
detail
SASH
wall
detail
window
detail

Magdiel

Owner-builders have been using formed masonry techniques since the 1870s when Goodrich and Fowler first popularized them. To a large extent each new method is an outgrowth of the one before it, incorporating slight improvements. Unfortunately, the costly time-consuming and laborious aspects of this work were passed on, too. An example of this is Goodrich's insistence that formwork should not be removed for at least forty-eight hours after pouring. Up to this day, this has remained unchallenged.

Presently, however, two of us have built formed masonry walls from which the formwork has been removed just *one hour* after it was cast. One hundred years ago form boards were made of two-inch-thick planks while today they are usually made of three-fourths-inch plywood. This difference, of course, is insignificant since any wooden form continues to be costly to build and awkward to handle. The forms we use, on the other hand, are metal, only three-sixteenths of an inch thick and easily fabricated at little cost. They are also easy to swing into place and to move while one works.

The breakthrough in forming poured masonry occurred in the southwest during the Depression days when Dan and John Magdiel patented their first Wall Building Machine. Dozens of poured concrete and stone buildings were subsequently erected by these brothers before the so-called Magdiel Form was perfected, manufactured and marketed. Unfortunately, the brothers failed to write about their method and so were unable to change popular beliefs about forming stonework.

The Magdiel Form is simply a thirteen-inch high by four-feet long metal container into which any masonry material can be dry packed. This same form has been used extensively for building rammed earth walls, for example. Dan Magdiel even built a house using a mixture of cow dung and bitumel (emulsified asphalt). For our purposes, the form works exceedingly well for building stone-cement walls. Building stone is placed against the form's metal sides and a fairly dry, gravelly mortar mix is packed between the stones. Immediately after packing, an ingenious release lever is pulled and the sides of the form separate, allowing its horizontal movement to the next four-foot long section to be worked. It is again clamped into position, and stone packing and mortar pouring is resumed. The complete form weights less than thirty pounds and one person can easily release, move and clamp it into a new position.

When the Magdiel brothers died in the early 60s, their Wall Building Machine, the Magdiel Form, passed into disuse. No one felt inclined to exercise public domain to the patent rights in order to re-manufacture the form. It is a device that sold more readily in the destitute days of the 30s than it would have during the affluence of the 60s and 70s. Furthermore, the Magdiel Form is more complicated to build than even the experienced do-it-yourself metal worker would care to tackle. The release mechanism, especially, requires machine shop precision fabrication.

Accordingly, we have preserved the basic principle of the Magdiel Form but have simplified it to the point where others can build their own form at minimal cost. We have also found that a horizontal building sequence which leap frogs from one section to

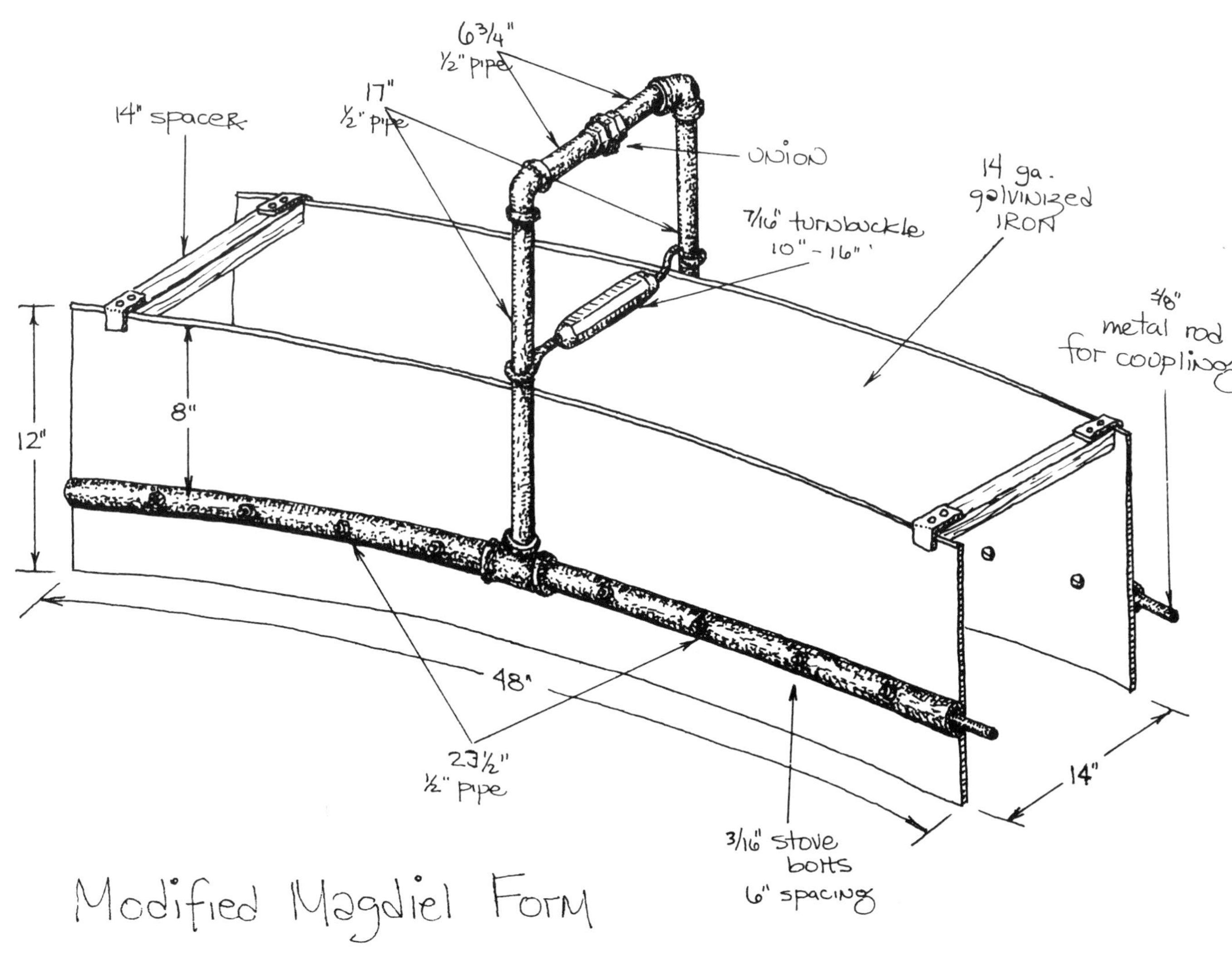

Modified Magdiel Form

the next is more adaptable for amateur construction *Two* forms that interlock are required in this action. As stone and mortar are packed into the second part of the form, sufficient time allows for the curing of the mass in the first form. When the second form is fully packed, enough time has elapsed for the first form to be removed and placed ahead of the second. In actual practice, one twelve-inch-high layer of wall is formed completely around the perimeter of the house before a second layer is begun. As many as three layers, amounting to three feet in height, can be formed in this manner in one day — depending on the length of the perimeter and attendant weather conditions. This dual forming method permits the use of a wetter mortar mix. The Magdiels used a fairly dry mix which they *tamped* around the packed stone. A wetter mix, however, *pours* easily and more readily around stone. If wall building is done during the cool of the year, it may be necessary to use a water-reducing set accelerator in your mortar mix. This is an amazing liquid which, when added to mortar, reduces the requirement for water by fifteen percent. It increases the compressive strength of mortar by three hundred percent in twenty-four hours. In effect, curing time is substantially reduced so that the forms can be removed sooner. Protex (1331 West Avenue, Denver CO 80223) is one company making an accelerating admixture called PDA High Early. Only fourteen ounces of this mixture is needed for each bag of cement.

Once you start laying stone in this form the relative simplicity and speed of this method compared to the laid and faced methods will become apparent. The first course of stone is begun directly on the footing. Lay a bed of mortar and then pack stone against both faces of the stone. Spaces between these face stones can be filled with rubble and mortar. One disadvantage of this method, however, is that as you lay the stones you cannot see how their faces fit against each other. Their fit can only

be judged by the shape of their building surfaces. It is, therefore, difficult to make conscious patterns or designs in the stonework.

Fill the form with stones until they project several inches above the top of the form. Be sure that all the stone you lay sits on sound bedding so none will push out once the form is moved along. Once both forms are filled, make sure the mortar in the first form is set; it needn't be hard, just firm. Then leap frog it over the second form, secure it in place and continue building.

When the first course is completed, clamp the forms on top of the finished section and begin building the next higher course. Make sure the forms are clamped, level and plumb. Continue in this manner until the top of the wall is reached.

Stones can be laid either dry, without mortar, or with a bed of mortar between them. Then, once the form is removed, the joints between the stones must be finished. If they have been laid in a bed of motar and the mud has not set hard, any excess mortar can be scratched away. Then a richer mortar mix can be made to fill the remaining space or the joint can remain recessed.

For the owner-builder who insists on the simplest, least costly, easiest to build, strongest and best insulated stone-formed housing, we recommend a fourteen-inch thick, curvilinear stone wall with a core of two-inch polyurethane foam. Inside and outside corner constructions are avoided by using the stone-formed technique, just as the amateur mason can also avoid building masonry headers, sills and high walls with this method.

Details

Steps

Although stone steps appear simple to build it is necessary to use care in their planning and construction. When you build steps it is vitally important to lay them out correctly. You must have a clear picture of how many steps are needed, the width of their tread and the height of their rise. It helps to make a detailed drawing of the proposed flight. Also test some steps with the rise and run you have selected, making sure they are not too steep or too shallow.

Steps are built sequentially, one upon the other. When the first step is being built, make the tread wide enough to accomodate the width of the riser of the next step. Continue building in this manner and, if you planned well, the last tread will be in position at the correct height. All steps should rise at the same rate. Even a small difference could cause an accident.

To build steps you will need a quantity of square-edged stone. The treads must have sharp, clean edges for good footing. Some masons use slate or similar smooth stone for treads, which must be flat and slightly forward sloping so that water and ice will not collect on them. Stone steps can be attractive and functional when built correctly but hazardous if built without care.

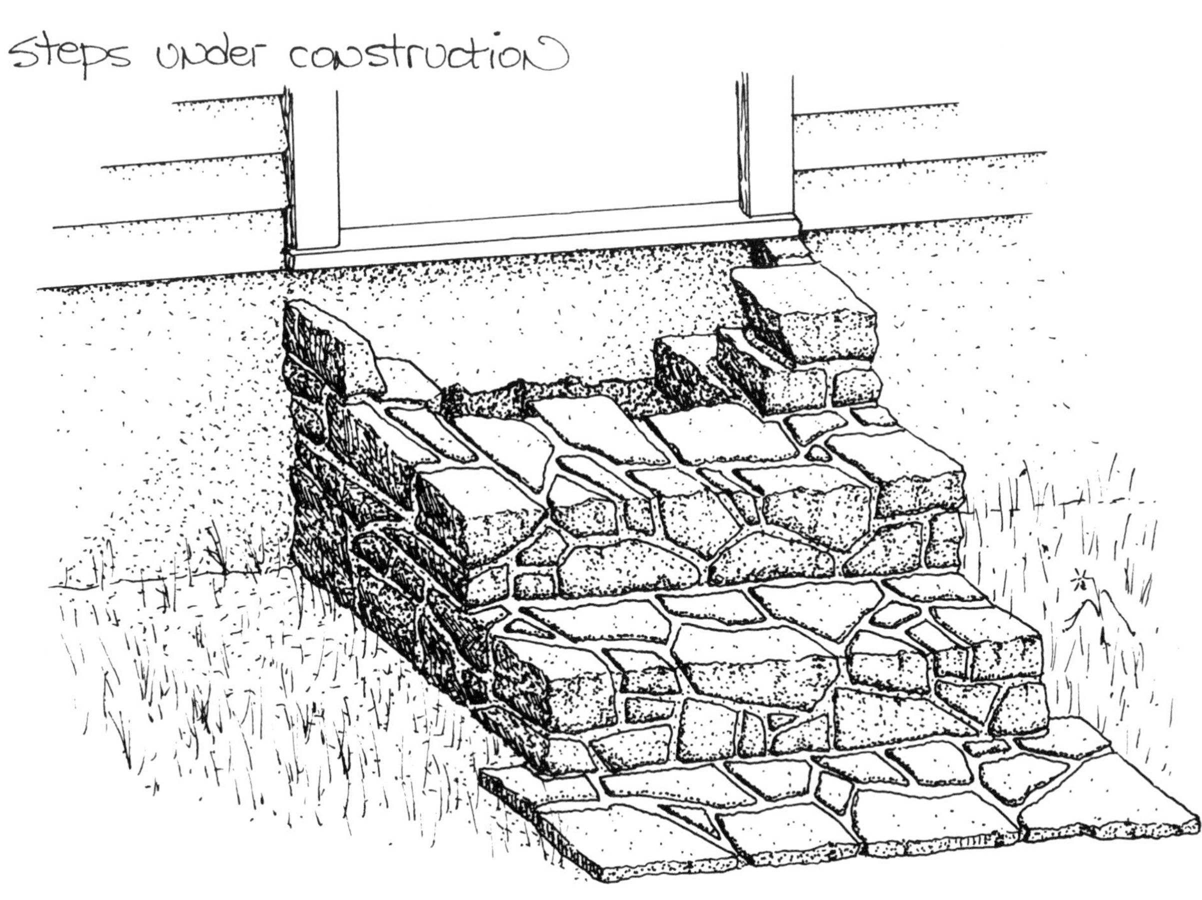
steps under construction

Floors

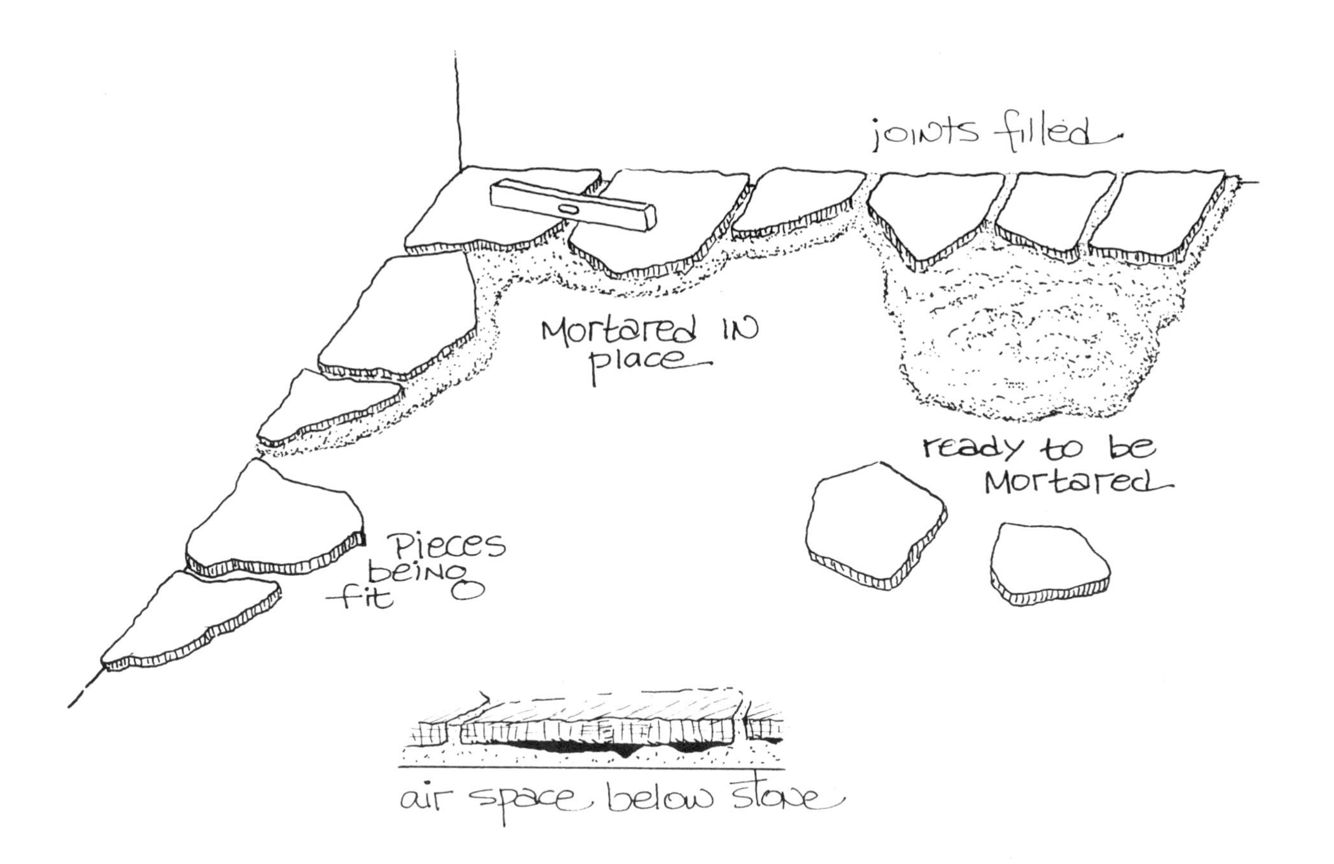

When one lays a stone floor or a hearth, the resulting surface should be flat and smooth. Stone is selected for its thinness and the flatness of its face. When laying a floor it is usually inconvenient to use pieces over three-inches thick. Slate, the traditional floor covering, has a hard surface and can be obtained in thin sheets. However, stone with similarly acceptable qualities can sometimes be found in fields and creek beds. It is also possible to find good flooring material among discards at the quarries which cut granite and marble for gravestones.

It is best to lay stone flooring on a solid surface, either directly on the ground or a concrete pad. Floors supported with wooden joists are not recommended even if these members are given extra support. It is impossible to keep a wood floor from flexing and cracking mortar.

When laying a stone floor the best place to start is at corners and around the edges. Position a number of pieces; then take them up, noting their arrangement. Trowel down several inches of mortar. It should be stiff for it will have to support the weight of stone above it. Do not smooth the mortar; let the stone mash it down. After you place the stone, tap it to the desired level with the handle of your hammer or a rubber mallet. Check this placement for level and adjust it. If the stone sinks too much it will have to be taken up to add more mortar. You may also have to use stone shims. Tap the top of the stone; if it sounds hollow then it is not seated firmly in its mortar bed. Work toward the center of the area you are covering. If it is large you may want to set guide lines to indicate level.

You can fill between stones with mortar as each is laid or after all are laid. It is best to fill joints level with the face of the stone so that no one will ever trip on an exposed edge. Building stone floors (on the horizontal plane), and hearths is much like building vertical walls.

Retaining Walls

Retaining walls must withstand forces that free-standing walls do not have to contend with. The purpose of a retaining wall is to stabilize a bank of earth. Dirt embankments tend to slide forward and level out. The reason for this is that surface water runs over these banks and, through the process of erosion, carries earth with it. Water and earth exert tremendous horizontal pressure on retaining walls. Although it helps to build a wall thick and strong, that alone will not withstand these forces.

To build a lasting retaining wall you must outwit nature. Pressure is relieved by constructing the wall to permit water to pass through. For this reason, the most effective retaining walls are built dry; they offer less resistance to flowing water. A mortared wall requires weep holes allowing the passage of water. A gravel drain field between the wall and the bank will allow water to flow freely around the ends of the wall.

Walls that curve inward offer less resistance to water than straight walls. A curving configuration is also stronger than a straight line. Build your retaining wall sloping inward rather than plumb; an earth bank tends to assume this shape naturally.

There are some retaining walls through which you do not want any water to flow. In this case, coat the surface interior of the wall with a waterproof material, such as asphalt or plastic — possibly both. Extra measures must be taken to let water pass around the ends or beneath this type of wall.

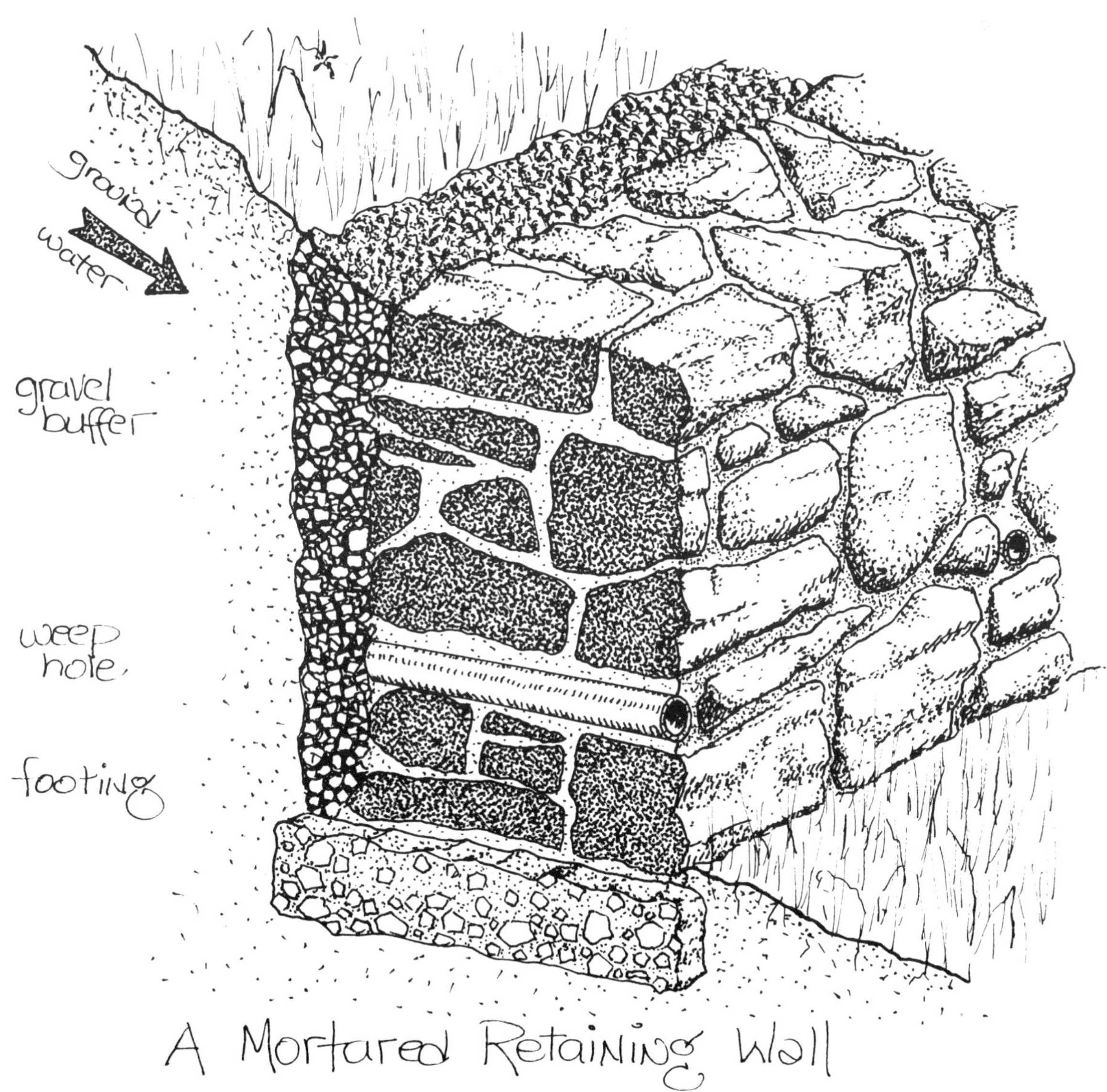

A Mortared Retaining Wall

Arches

An arch is a collection of stones working together to span an opening that is wider than any one of them. A finished arch needs the support of neither wood or metal. Like any other stone construction the force that makes an arch work is gravity. Gravity wedges each stone so tightly that it cannot move. More than any other stone structure, an arch must be built carefully of closely fitted pieces.

To build an arch you must first assemble a form to support the stones until they are able to support themselves as a unit. Select all stones for the arch and place them, dry, on the form. After they have proven to fit well together they can be mortared into place. Once the mortar is set, the form can be taken down and the arch will be self-supporting.

Arches exert outward pressure on walls supporting them. These walls must be strong enough to withstand this pressure. Arches can be any number of shapes, from the tall gothic style to a straight line. The flatter the arch the more outward pressure it will exert on the supporting walls.

Each stone in an arch must be shaped so that it fits snugly against its neighbor on either side. Although the keystone is traditionally the symbol of the strength of an arch, it is no more important than any other stone therein. Appropriately shaped stone for arches is difficult to find. In most cases the mason must shape them to fit. Each piece should be wider at its top than at its base. Its sides must radiate symetrically outward from a point at the center (the focal point). These shapes can be defined with the use of a radius string attached to this central point. You must build your arch with judgement and confidence so that when the form is finally removed you will not be afraid to pass beneath it.

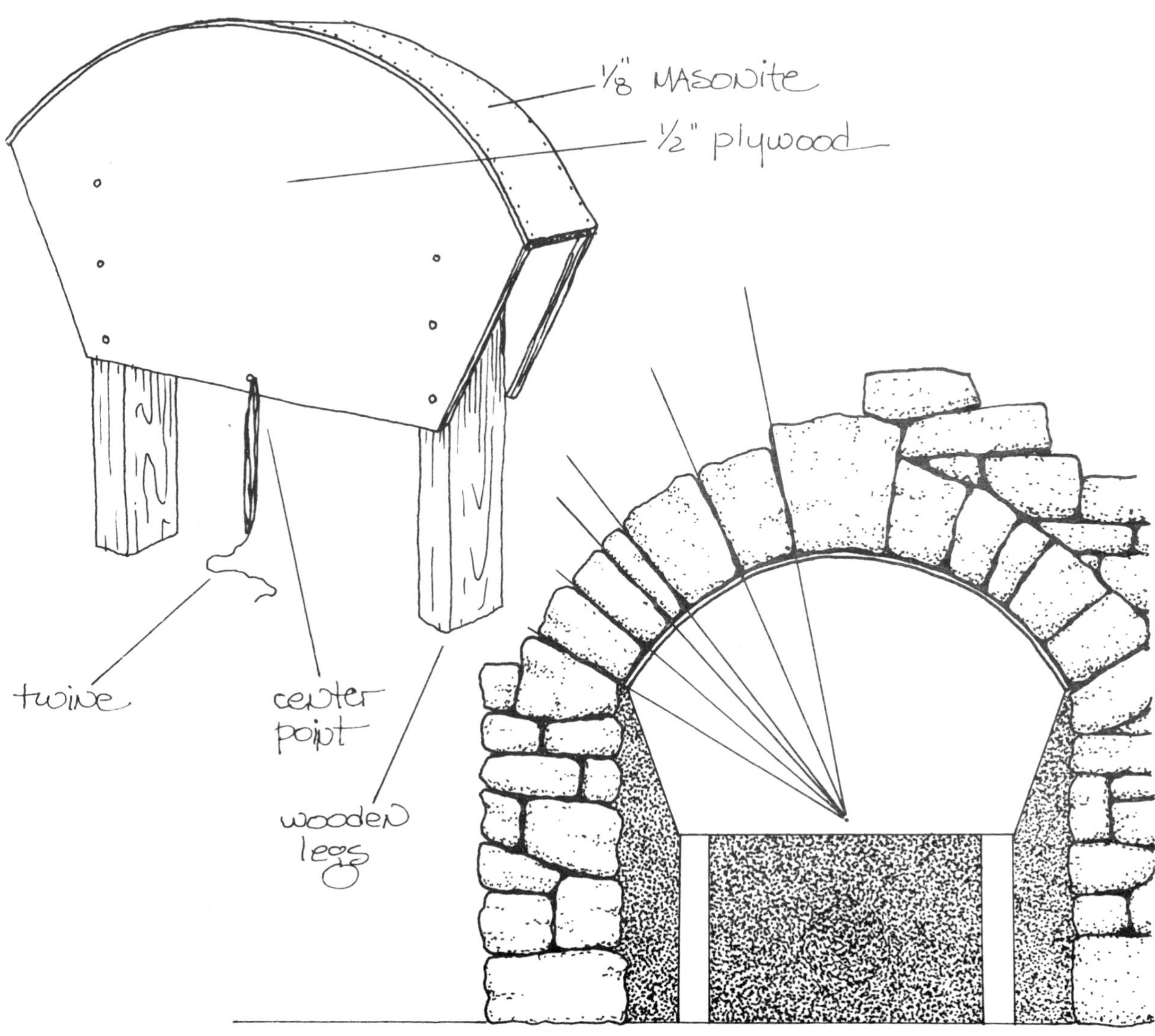
1/8" MASONITE
1/2" plywood
twine
center point
wooden legs

Fireplaces

A fireplace is the ultimate symbol of warmth, comfort and security. Indeed, there is something mystical about the human response to the combination of stone and fire. In the past fireplaces were built entirely of this material. Intense heat, however, is probably the most destructive force to which the stone in fireplaces may be subjected. Look into the heat chamber of most old fireplaces and you will likely find that the stone therein is cracked and crumbling.

Contemporary stone fireplaces are built with this fact in mind. Their construction includes a variety of heat-resistant materials to insure a sound and safe fireplace and chimney. The heat chamber is lined with firebrick or metal, a damper is installed to regulate air flow and the chimney is lined with ceramic flue tile. Stone and mortar provide structural support and an attractive facade for these inner workings.

Building a fireplace and chimney is a complex process. There are a wide variety of basic designs from which to choose and hundreds of variables to be considered for each design. It would be impossible to adequately include all of the details of fireplace design and construction in this book.

The diagrams provided here are meant only to give the reader a general picture of how a fireplace and chimney are put together. They illustrate some of the construction details needed for laying stone in a fireplace, but much additional information will be required if you want to build your own.*

*In preparation: *The Owner-Builder's Guide to Fireplace Construction*. Write the publishers to be placed on the mailing list.

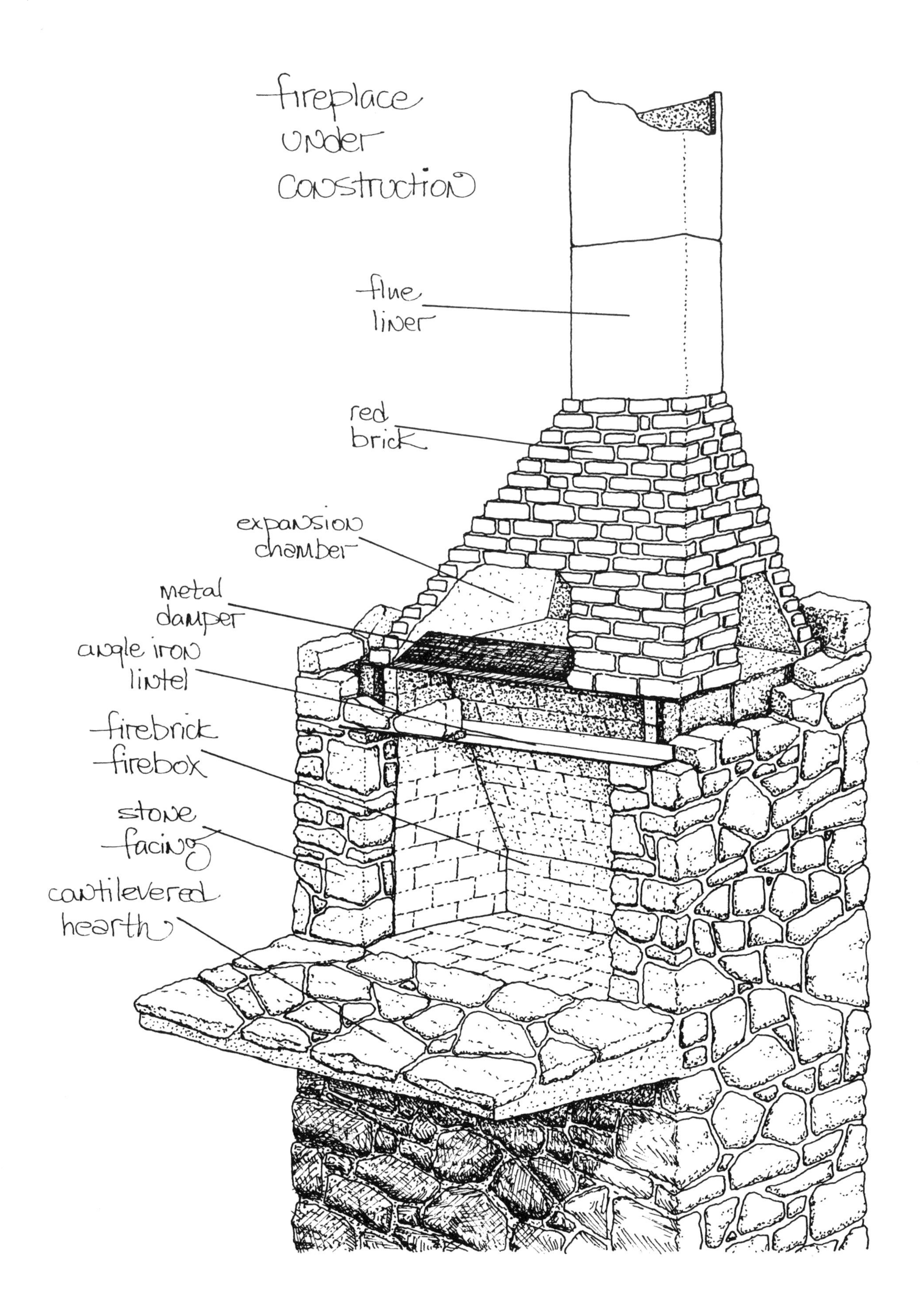
fireplace
under
construction
flue
liner
red
brick
expansion
chamber
metal
damper
angle iron
lintel
firebrick
firebox
stone
facing
cantilevered
hearth

Scaffolding

If the plans for your house include stone walls or a chimney you will need to use scaffolding. Its quality will make a big difference in the ease with which your work progresses. Take time to construct adequate, well-braced scaffolding. Working height should be between knee and chest level. When you have to lay stone higher than that, it is time to raise the scaffold boards. Good scaffolding provides plenty of room to stand as well as adequate space to stockpile stone. There should also be a board at waist height on which to place tools and a mortar tub. Put a safety rail, a strong one, along the outside edge of the decking. The necessity for careful scaffold building cannot be overemphasized. Every day you will have as much as two tons of rock and cement besides yourself on this staging. If it should give way you might find yourself on your slab in a lot of trouble.

Scaffolding can be provided in several ways. The builder can construct his own using surplus framing lumber. This is the least expensive but not necessarily the easiest method. Tubular metal scaffolding can be rented in most areas at very reasonable rates. This scaffolding is designed to meet the specifications mentioned above. For masonry, specify the type with ladders on the side so that the scaffold boards can be raised at two-foot increments. That way your work will always be at a comfortable height. Jacks are useful for leveling the scaffold on sloping or irregular ground.

If you are doing a lot of work requiring scaffolding you might decide to purchase some. One simple and convenient variety is the pump jack. The Hoitsma Adjustable Scaffold Bracket Company (Box 452, River Street Station, Paterson NJ 07525) manufacture a mason's pump jack that is well-suited for stone facing work. A foot-operated lever raises or lowers the scaffold to any convenient working height. Whichever method of scaffolding you use, be sure it is built strong enough to hold the weight of mason, stone and mortar.

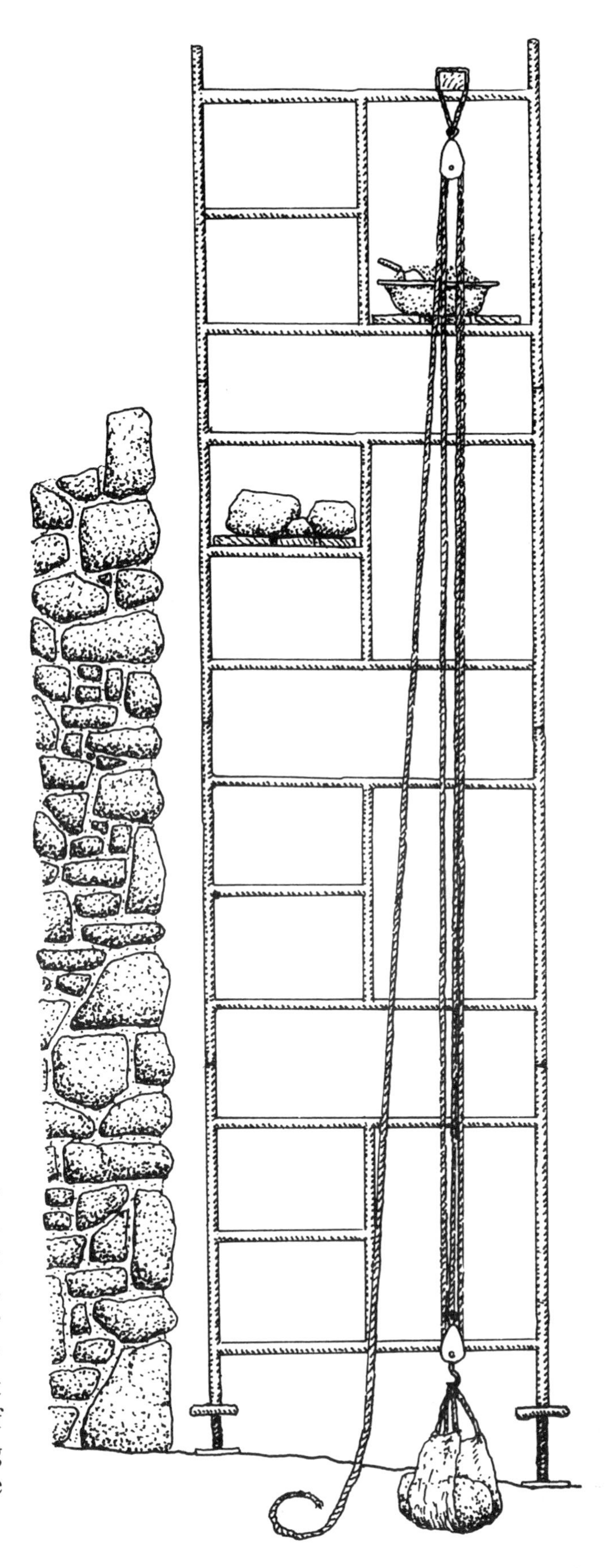

Glossary

Accelerator — A substance added in small quantities to concrete or mortar to hasten its hardening rate. Calcium Chloride is often used for this purpose. Accelerators are useful when working in cold weather to make the cement set before it freezes.

Admixture — A substance added in small amounts to concrete or mortar to alter its properties. Admixtures are used as accelerators, plasticizers and air-entraining agents.

Aggregate — Stone, gravel, sand or any similar inert material which is bound together with cement to make mortar or concrete. The aggregate composes the bulk and compressive strength of the mix.

Anchor — Any means used to mechanically bind a masonry mass to a foundation or wall. Generally made of metal, anchors come in a variety of styles from flat corrugated wall ties to "Z" bars made of round stock.

Arch — A curved masonry structure which spans an opening without other support. Stone arches are composed of units all smaller than the opening itself but wedged together to form a continuous bridge.

Arch stone — (Voussior) Any one of the wedge-shaped pieces in the arch.

Ashlar — 1. A stone with a square or rectangular face. 2. The style in which this stone is laid.

Backfill — Earth or stone used to fill behind a foundation or retaining wall. Backfill takes time to compact and should not be relied on to bear weight.

Backing — 1. The surface against which facing stone is laid. 2. Concrete or stone used to fill behind the face in a wall.

Basalt — A dense textured, igneous rock relatively high in iron and magnesia minerals but relatively low in silica. Basalt is generally dark gray to black and feldspathic.

Batch — One mixing of concrete or mortar.

Batterboard — Fixed horizontal boards located at the outside of foundation corners. Nails are set in the top edge of these boards and connected with lines to indicate excavation, footings and floor level.

Bed — 1. A prepared surface on which stone is laid. 2. The surface of a stone parallel to its stratification.

Bed width —The thickness of a faced wall.

Bedding — A layer of mortar upon which stone rests.

Binder — Any powdery substance mixed with aggregate to form mortar or concrete.

Bond — 1. A regular pattern according to which stone is laid in a wall. 2. The adhering of mortar to stone.

Bondstone — (Bonder) Inlaid masonry, a large flat stone used to unify the two sides of a wall. When veneering it is a stone laid flatways which anchors the wall to the backing. It is not needed when metal ties are used.

Bridge stone — A flat stone spanning an opening or gap.

Brownstone — A sandstone of brown or redish-brown color. This hue comes from a prominant amount of iron oxide.

Building inspector — A state or county employee whose duty it is to enforce the building code. The code prescribes where and how one may build his own house as interpreted by the inspector who often knows nothing about how the owner-builder can best solve his housing needs.

Building surface — Any surface on a stone used to support weight in a wall.

Bush hammer — A hammer whose face is composed of a number of points, often used to smooth faces on soft stone like limestone, marble and sandstone.

Cap stones — The uppermost, and often decorative stones used to finish a wall.

Carve — Shaping stone by cutting a design or form, the trade of a sculptor.

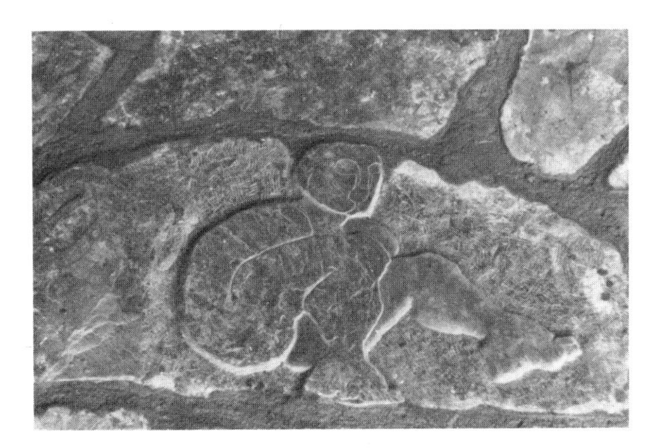

Cast stone — Imitation stone.

Cavity wall — Any hollow wall. The two sides are separated by a continuous air space and connected by wall ties.

Cement — A binder (such as portland cement) which is mixed with aggregate to form concrete or mortar.

Cleavage — The ability of a rock to break along a natural surface; the surface of this natural break.

Chinker — A small stone used to fill gaps between larger ones in a wall.

Clay mortar — A mixture of clay and water used to fill gaps between stone in a wall.

Cobblestone — A naturally rounded stone large enough for paving. This term is now also used to describe any paving block.

Concrete — A mixture of water, sand, gravel and a binder (portland cement) which hardens into a stone-like mass.

Corbel — To lay stone so that it projects from the surface of a wall. Corbelled stones are often used to support beams.

Course — A horizontal layer of stone extending the length of a wall.

Cut stone — Any stone cut, chipped or machined to a given size, dimension or shape.

Damp proofing — The coating a wall, above grade, with a compound that is impervious to water.

Damper — A metal plate in the flue of a chimney used to regulate the draft.

Dormer — A vertical opening in a sloping roof which is usually provided with its own pitched covering.

Dressed stone — Stone that has been squared all around and smoothed on the face.

Dry wall — A stone wall that has been built without the use of mortar.

Efflorescence — A crystaline deposit appearing on stone surfaces that is caused by soluble salts carried out of the stone by moisture. It usually comes from mortar or concrete backing. It's harmless.

Eyeball —A check for plumb by lining up the wall and the string on a plumb bob held at arm's length.

Face — The exposed portion of a stone in a wall.

Fat mortar — (Rich mortar) A mix with more than the usual amount of cement, lime or other binder. It is used when a stickier consistancy is desired. Fat mortar is not necessarily stronger than a regular mix and, in fact, is often brittle.

Fieldstone — Loose pieces separated from ledges by natural processes and scattered on the ground.

Fireclay — A binder capable of withstanding extreme heat without disintegrating. It is also added to mortar to make it more plastic.

Flagstone — Thin slabs of stone used for paving (flagging) walks and patios.

Footing — The support upon which the foundation wall sits.

Foundation — The whole masonry support for a building.

Freemason — A term from the Middle Ages referring to a skilled mason who is capable of cutting freestone.

Freestone — A stone that may be cut freely in any direction without fracture or splitting.

Furring — A cavity within an exterior wall providing space for insulation and a vapor barrier.

Gauge — The proportion of different materials in mortar or concrete.

Grain — The plane along which a stone splits.

Granite — A fine to coarse-grained igneous rock formed by volcanic action. Granite is a hard stone and difficult to shape.

Green mortar — Portland cement mortar after its initial set but before it has begun to harden properly. Mortar is green in color for about a week in cold weather.

Grout — Mortar of pouring consistancy.

Hearth — In a fireplace the area directly in front of the fire chamber. A masonry hearth is important to prevent fires caused by flying sparks.

Igneous — One of the three great classes of rock. Igneous rock is solidified from a molten state as, for example, granite or lava.

Insulation — Any material used to prevent heat or cold from passing through a wall.

Jack arch — One having horizontal or nearly horizontal upper and lower surfaces, also called a flat or straight arch. The less curve in an arch, the more outward pressure it puts on its supports.

Joint — The space between two stones in a wall.

Keystone — The last wedge-shaped stone placed in the crown of an arch. Although it symbolizes completion, this stone is no more important than other arch stones.

Lean mortar — A mix using less binder than customary. It is usually difficult to spread.

Level — (spirit level) A straight-edged tool used to determine true horizontal and vertical planes by means of curved glass vials containing liquid and air.

Lime — Chalk and other forms of calcium carbonate burnt in a kiln to powdery consistancy. It is called quick lime until it is soaked in water when it then becomes hydrated or slaked lime.

Lime mortar — Lime and sand mixed with water. Lime mortar has a plastic consistancy, making it easy to spread, but is soluble in water. Lime is also added to portland cement mortar to make it more plastic.

Limestone — A sedimentary rock composed largely of calcium carbonate. This is a soft and workable stone, often used for carving.

Line — A string, usually made of nylon, used for setting up building work.

Line level — A small spirit level which is suspended in the center of a taut line to compare points over a distance. It is not as accurate as a water level or transit.

Line pins — Metal pins about three inches in length. They are inserted in the spaces between stone in a wall and used to hold guidelines.

Lintel — A single piece of metal, wood or stone used to span an opening.

Marble — A metamorphic rock composed essentially of calcite and dolomite, generally a recrystalization of limestone.

Mash hammer — A Scottish term for a small double-faced club hammer or sledge hammer weighing two to four pounds.

Masonry — Stacked construction often set in mortar.

Mortar — Sand and water mixed with a binder and used to fill gaps between masonry units.

Metamorphic — A class of rock which has been changed or altered by external agents, such as deep-seated heat or pressure.

Mosaic — A style of stone laying which is generally irregular with no definite pattern. Generally, the stone used is of no specific geometric shape.

Mud — Mason's term for mortar.

Natural bed — The surface of a stone parallel to plane at which it formed in the ground.

Natural Cleft — This generally refers to stones which were formed in layers. When these stones are cleaved or separated along a natural seam, the resulting surface is called the natural cleft surface.

Non-staining mortar — Mortar composed of materials which will not stain the surface of stone. It usually has a very low alkali content.

Parapet wall — The part of any wall which is entirely above the roof line.

Parging — Coating a masonry wall with a layer of mortar. This is done either to fill gaps, smooth surfaces or damp proof.

Pit-run gravel — Naturally occuring aggregate excavated from a pit.

Plastic — A term describing sticky, workable mortar.

Plug — A small, pointed wooden peg pushed into a hole in a wall where a screw, nail or other fastening device can later be secured.

Plumb line — A string on which a weight (bob) is hung stretching it in a vertical direction. If the string is braided, like fishing line, the weight will not spin.

Pointing — The final filling and finishing of mortar joints.

Polyurethane foam — A plastic foam available in sheets with the following properties: high ratio of strength to weight, low thermal and acoustical conductivity, low transmission of water vapor and high dimensional stability.

Portland cement — A binder used in mortar and concrete which sets with stone-like qualities.

Projections — Stones jutting from a wall giving the effect of ruggedness.

Quarry — An operation where a natural deposit of stone is removed from the ground.

Quarry face — (quarry dressed) A description of stone as it comes from the quarry. It is squared on the sides but has a rough face.

Quarry sap — The moisture in stone freshly cut from the quarry. Some stone, like slate, is soft when first quarried but hardens when it dries out.

Quartzite — A compact granular rock composed of quartz crystals, usually firmly cemented into a homogeneous mass. Its compressive and tensile strength are extremely high; its color range is wide.

Quoins — Stones at the corner of a wall which are emphasized by their contrasting size, projection or different finish.

Raked joint — A mortar joint in which the mortar has been scraped back with a tucking trowel or similar tool. This is done either in preparation for pointing or as a finish itself.

Random rubble — A masonry style in which the stone is of irregular shape and size, not laid in courses.

Recessed joint — A mortar joint in which the pointing is set back from the face of the stone, creating a strong shadow effect. With this style there is less danger of the mortar cracking out than with flush joints.

Reinforcing rod — (rebar) Round steel rods available in twenty-foot lengths and three-eights to one-inch diameters, used for increasing the tensile strength of concrete.

Ready mix — Concrete mixed at a central batch plant and delivered to the site in mixer-equipped trucks.

Return head — A cornerstone having a right angle and two sides suitable as faces.

Reveal — The distance from the face of a wall to a window or door set back in the wall.

Rich mortar — See fat mortar.

Rift — The indication on a stone's surface of the plane along which it will split.

Riprap — Irregular-shaped stone used for facing embankments and fills. Stone thrown together without order to form a foundation or sustaining wall.

Rock — An integral part of the earth's crust composed of an aggregate of minerals.

Rubble — Building stones which are not smoothed to give fine joints, like ashlars, but are sometimes squared and laid in courses.

Saddle — A flat strip of stone projecting above the floor between the jambs of a door; a threshold.

Sandstone — A sedimentary rock consisting usually of quartz cemented with silica, iron oxide or calcium carbonate. Sandstone is durable, has high crushing and tensile strength and a wide range of colors and textures.

Scaffolding — A temporary steel or timber erection for supporting people and materials during building.

Schist — A metamorphic rock composed predominently of minerals whose long dimensions are oriented in approximately parallel positions or planes of foliation. Because of this structure schists split readily along these planes. The most common schists are composed of micas and generally contain subordinate quartz and feldspar of fine-grained texture. All graduations of texture exist between schist and gneiss, a more coarsly foliated rock.

Scoria — Irregular masses of lava resembling clinkers or slag. It may be cellular, dark colored and heavy.

Sedimentary — A class of rock formed of sand, mud or clay deposited in layers in the ocean.

Serpentine — A hydrous magnesium silicate material of igneous origin, generally a very dark green color with markings of white, light green or black. One of the hardest varieties of natural building stone.

Set — The initial hardening of concrete or mortar, a chemical process.

Shear — A type of twisting stress poorly countered by unreinforced masonry. A wall is in shear when it is subjected to a pair of equal forces which are opposite in their direction.

Sill — A flat stone used under windows, doors and other masonry openings.

Slab — A lengthwise cut of a large quarry block of stone.

Slate — A very fine-grained metamorphic rock derived from sedimentary shale. Characterized by an excellent parallel cleavage entirely independent of original bedding, this rock may be split into relatively thin slabs.

Soapstone — A variety of talc with a soapy or greasy feel. It is used primarily for hearths and fire chambers and is known for its stain-proof qualities.

Spall — A stone fragment that has been split or broken off. Spalls are primarily used for filling gaps in rough masonry.

Stone — Sometimes synonymous with rock but more properly applied to individual blocks or fragments taken from the mother mass. Stone is the building material obtained from rock.

Stone boat — A wooden or steel tray mounted on sledge runners for hauling stone short distances over a rough road or trail.

Stratification — A formation produced by deposition of sediments in beds or layers (strata).

Strip rubble — Generally speaking strip rubble comes from a ledge quarry. The beds of stone are uniformly straight just as they are removed from the ledge.

Tailings — Stone or earth refuse from a mining operation.

Tamp — To compact mortar, earth or gravel by repeated pounding with a heavy weight.

Through stone — A bond stone which is seen on both faces of a wall.

Tooled finish — Parallel grooves on the face of a stone cut four, six or eight to an inch.

Toothed chisel — A stone cutting tool used for fine work in shaping soft stone.

Vapor barrier — An airtight skin placed in a wall to prevent the penetration of moisture.

Veneer — Stone used as decorative facing which is not meant to be load bearing.

Wall tie — A metal anchor used to bind masonry to a wall.

Water table — A projection of lower masonry on the outside of the wall slightly above the ground, preventing the upward movement of ground water.

Wedging — Splitting of stone by driving wedges into planes of weakness.

Weepholes — Small gaps left in the joints at the foot of a wall to allow water to pass through.

Z-Bar —A Z-shaped metal rod used as an anchor.

CONSTRUCTION PROPERTIES OF BUILDING ROCK

Geologic origin	*Physical type*	*Rock name*	*Water absorption resistance*	*Insulative quality*	*Mechanical strength*	*Durability*	*Surface character*	*Presence of impurities*
Igneous (formed from molten material)	Intrusive coarse grained	Granite	Good	Fair	Good	Good	Good	Possible
		Diorite	Good	Fair	Good	Good	Good	Possible
	Extrusive fine grained	Basalt	Excellent	Excellent	Good	Good	Good	Seldom
		Obsidian	Excellent	Good	Good	Good	Good	Possible
Sedimentary (sediments deposited by wind and in water)	Calcareous calcite	Dolomite	Fair	Good	Good	Fair	Good	Possible
		Limestone	Poor	Poor	Good	Fair	Good	Possible
	Siliceous silica	Shale	Poor	Good	Poor	Poor	Good	Possible
		Sandstone	Fair	Good	Fair	Fair	Good	Seldom
		Chert	Poor	Fair	Good	Poor	Fair	Likely
		Conglomer	Poor	Fair	Fair	Fair	Good	Possible
Metamorphic (prolonged heat and/or pressure)	Foliated parallel layered	Slate	Excellent	Good	Good	Good	Poor	Seldom
		Schist	Good	Good	Good	Good	Good	Seldom
	Nonfoliated	Quartzite	Excellent	Poor	Good	Good	Good	Seldom
		Marble	Excellent	Poor	Fair	Good	Good	Possible
		Serpentine	Good	Fair	Fair	Fair	Poor	Possible

Credits

Photos

Robert Agoston: 6, 14, 34, 65, 75, 79 rt, 81, 115, 117 bot, 125, 176 bot.

Brandi: 168.

Bob Brooks: 41 bot, 64, 126, 130, 133.

Paul Corey: 146, 148, 149, 150, 151.

Robert Dorksen: 23 bot.

Ed Eberle: 135, 143.

Ernest Flagg, *Small Homes*: 137, 138, 140, 154.

Greg Flint: 24.

Fresno Bee: 8.

Richard & Sandi Fryer: 154, 155.

Tate Hamlet: 47 tp rt, 51 tp, 60, 182 tp.

Susanna Holzman: cover bot, 50, 56 bot, 71, 87, 103 lft, 181 bot rt.

Heidi Kern: 1.

Joe Kern: 21 bot.

Ken Kern: 17, 19, 20 bot, 36 tp, 37 bot, 41 tp, 46 tp rt, 48 bot, 49 rt, 57 bot, 68, 80 bot lft, 94, 118, 127, 128, 131 tp rt, 134, 156, 158, 159, 160, 182 bot.

Steve Magers: 21 tp, 25 bot rt, 26, 28, 29, 31, 38, 42, 47 tp lft, 49 lft, 53 bot, 61, 63, 67, 80 tp lft, 80 lft, 82, 92, 105 bot, 119 tp, 120, 161, 165, 171, 173.

Bill Marriott: back cover, 2, 10, 18, 20 tp, 25 tp, 32, 46 tp lft, 46 bot, 47 bot rt, 48 tp lft, 51 bot, 53 tp, 57 tp, 58, 69, 72, 105 tp rt, 175, 180, 181 bot lft.

Leslie & Terry McMenamin: 36 bot, 105 tp lft, 170, 174.

Scott & Helen Nearing: 144, 145 bot.

Lou Penfield: cover tp, 22, 23 tp, 25 bot lft, 33 tp, 33 bot lft, 35 tp, 37 tp, 44, 45, 55, 85, 100, 101, 102, 103 rt, 104, 106, 107, 108, 109, 110, 112, 114, 116, 117 tp, 119 bot, 122, 131 tp lft, 162, 164, 166, 167, 172, 176 tp, 177 bot, 178, 181 tp rt.

Tisa Penfield: 79 lft, 84, 90, 93, 131 bot.

Frazier Peters, *Pour Yourself a House*: 142.

John Rodgers: 35 bot, 74, 91, 111, 124, 132.

Robert Roskin: 12, 13, 39, 47 bot lft, 177 tp lft.

Karl Schwenke: 145 tp.

Scope Photographers: 76.

Lynn Walters: 9, 73, 77, 88, 177 tp rt.

Lewis & Sharon Watson: 152, 153.

Drawings

Steve Magers

Typing & Editing

Barbara Kern